CDL PRACTICE TESTS

INCLUDES FULL LENGTH EXAMS FOR ALL CLASSES
+ AUDIOBOOK, FLASHCARDS AND ONLINE VIDEOS

LEARNIK© PRESS

Third Edition
First Edition: February 2022
This Paperback Edition First Published in January 2024

TABLE OF CONTENTS

INTRODUCTION

In order to operate any commercial motor vehicle such as dump trucks, tractor trailers, passenger buses, and the likes, you will first need to get a commercial driver's license. Taking your CDL test is the basic step to starting your professional truck driving career and you will need to study a comprehensive handbook before doing so. Although this book contains only practice tests and multiple-choice answers, you can check out the **"CDL Study Guide 2025 - 2026"** for detailed information about acquiring your commercial driver's license.

In the pages below are a total of 378 questions which cover all the topics you need to know to pass your CDL tests. Included are practice tests on Hazardous materials, School Buses, Doubles/ Triples, Pre-trip inspections, Tanker vehicles, Air Brakes, and Combination vehicles which will help you prepare for your tests adequately. Get started now and obtain your Class A, B or C CDL license in no time!

As you progress through this workbook, be sure to visit page 48, where you'll find detailed instructions on how to access our valuable bonus materials via the 'Learnik' e-learning platform, enhancing your study experience with additional resources.

1_ General Knowledge Practice Test

This full-length CDL General Knowledge practice test contains 43 questions. This test contains more CDL General Knowledge practice questions than you'll probably need to pass the General Knowledge endorsement, but why take any chances? Answering 43 practice questions will prepare you better than answering a mere 10 questions!

1. Which of the following statements about an escape ramp is false?

- ☐ **A)** It turns uphill
- ☐ **B)** It has no use when one is traveling fast
- ☐ **C)** It is made of soft gravel
- ☐ **D)** It is located a few miles from the top of a downgrade

2. How many tie downs should you use for a 20-foot load?

- ☐ **A)** 1
- ☐ **B)** 2
- ☐ **C)** 4
- ☐ **D)** 3

3. What is the best way to calculate one's following distance in seconds?

- ☐ **A)** Use a digital stopwatch to time the interval between your vehicle and the vehicle ahead as it passes a specific landmark.
- ☐ **B)** Observe when the rear of the vehicle ahead passes a fixed point, such as a sign or a tree, and then count the seconds until your vehicle reaches the same point.
- ☐ **C)** Choose a fixed point that the vehicle in front passes, and then count "one-thousand-one, one-thousand-two..." until you reach the same point.
- ☐ **D)** Measure the time it takes for your vehicle to pass two fixed points on the road, ensuring a safe following distance from the vehicle ahead.

4. How can one start moving without rolling back?

- ☐ **A)** Put on the parking brake
- ☐ **B)** Use the trailer brake hand valve
- ☐ **C)** Engage the clutch before removing your feet from the brake
- ☐ **D)** All of the above

5. In a pre-trip test with your instructor while examining hoses, what will you have to look out for?

- ☐ **A)** The location of a dipstick
- ☐ **B)** Frays in the water pump
- ☐ **C)** Puddles on the ground
- ☐ **D)** Windshield water fluid level

6. The starter key should be in your pocket key while performing the pre-trip test inspection because...

- ☐ **A)** So no one can move the vehicle
- ☐ **B)** The truck could get stolen
- ☐ **C)** Someone might start the truck
- ☐ **D)** Any and all of the above could happen

7. One would lose their CDL driving privileges if they are convicted of a second DUI offense in a private or commercial vehicle for how long?

- ☐ **A)** For life
- ☐ **B)** At least a year
- ☐ **C)** 10 years
- ☐ **D)** 1 year

8. One should not use a B:C fire extinguisher on which type of fire?

- ☐ **A)** Gasoline
- ☐ **B)** Wood
- ☐ **C)** Electrical
- ☐ **D)** Grease

9. The following tasks are involved in the checking and inspecting of a cab except...

- ☐ **A)** Starting the engine and putting it in neutral
- ☐ **B)** Checking on the air pressure gauge
- ☐ **C)** Transmission control check
- ☐ **D)** Listening for unusual noise

10. A normal tractor-trailer takes how many seconds to clear a double track?

- ☐ **A)** More than 15
- ☐ **B)** More than 30
- ☐ **C)** 14
- ☐ **D)** 15

11. What is black ice?

- ☐ **A)** A thin layer of clear ice
- ☐ **B)** A mixture of rain and snow
- ☐ **C)** Dirty snow
- ☐ **D)** Ice with dark substances in it

12. How can one prevent another accident from happening in a place where an accident has already occurred?

- ☐ **A)** Drink in order to calm your nerves
- ☐ **B)** Get to the higher ground
- ☐ **C)** Stay in your vehicle and wait for help to come
- ☐ **D)** Put warning devices so as to make sure that other vehicles are aware

13. How many hours of sleep does the average person need per night?

- ☐ **A)** 7-8 hours
- ☐ **B)** 8-9 hours
- ☐ **C)** 6- 8 hours
- ☐ **D)** 6-7 hours

14. Is it absolutely safe to remove a radio cap if the vehicle engine is not overheated?

- ☐ **A)** Yes
- ☐ **B)** Yes, as long as there is no overflow
- ☐ **C)** As long as the radiator is not damaged, yes
- ☐ **D)** No

15. What will help a driver get sober?

- ☐ **A)** Fresh air
- ☐ **B)** Time
- ☐ **C)** Coffee
- ☐ **D)** A glass of water

16. Which will get stuck at a railroad crossing?

- ☐ **A)** A car carrier
- ☐ **B)** A moving van
- ☐ **C)** The lowboy
- ☐ **D)** All of the above

17. When driving on a wet road, one should reduce their speed by...

- ☐ **A)** One-third
- ☐ **B)** 60 percent
- ☐ **C)** One-quarter
- ☐ **D)** One-half

18. In what situation should one totally downshift?

- ☐ **A)** When starting up a hill and finishing a curve
- ☐ **B)** When starting down a hill and entering a curve
- ☐ **C)** When starting down a hill and finishing a curve
- ☐ **D)** When starting up a hill and entering a curve

19. Which of the following is not a benefit of making plans when you spot a hazard?

- ☐ **A)** To avoid having to perform an action like sudden braking that is much more likely to cause an accident
- ☐ **B)** Improving the safety of other drivers on the road as well as your own by being prepared
- ☐ **C)** Gives you more time to hang up your phone call or finish something you might have been eating or drinking
- ☐ **D)** Gives you more time to do something

20. Where should you be looking ahead of your vehicle while driving?

- ☐ **A)** Straightforward at all times
- ☐ **B)** Toward the left side of the roadway
- ☐ **C)** Toward the right side of the roadway
- ☐ **D)** Both far and near, then back and forth

21. What is the greatest amount of play that should be in your steering wheel?

- ☐ **A)** 1 inch / 10 degrees
- ☐ **B)** 5 inches / 5 degrees
- ☐ **C)** 2 inches / 10 degrees
- ☐ **D)** All of the above

22. What is the major difference between road rage and aggressive driving?

- ☐ **A)** Intent to do harm or physically assault
- ☐ **B)** Level of anger
- ☐ **C)** Outcome of the situation
- ☐ **D)** All of the above

23. What action should be taken when your vehicle begins to hydroplane?

- ☐ **A)** Accelerate very slightly
- ☐ **B)** Brake gently to start slowing your vehicle
- ☐ **C)** Release the accelerator
- ☐ **D)** Let go of the wheel so it can correct itself again

24. Why is it a good idea for drivers to carry out the pre-trip inspection exactly the same way every time?

- ☐ **A)** Most companies require you to do so
- ☐ **B)** You will be less likely to forget something
- ☐ **C)** You should not do it the same way every time as doing so may make you too comfortable in your inspection
- ☐ **D)** The law requires you to do it the same way every time

25. What is the most important and obvious reason to always conduct a pre-trip inspection of your commercial motor vehicle?

- ☐ **A)** For safety reasons
- ☐ **B)** To make sure your paint is not chipped
- ☐ **C)** Employer liability
- ☐ **D)** To see if your vehicle needs a wash

26. Which of the following are correct methods for knowing when you should shift up?

- ☐ **A)** The road speed and engine speed
- ☐ **B)** Wind speed and shift indicator light
- ☐ **C)** Every 3 to 5 seconds and road speed
- ☐ **D)** None of the above

27. When stopped on a hill, what is the proper procedure to start moving without rolling back?

☐ **A)** Engage the clutch before you take your right foot off the brake

☐ **B)** Put blocks behind the tires

☐ **C)** Increase engine rpm and let the clutch out quickly

☐ **D)** Leave the trailer brakes on and drag the tires, releasing the brake after moving forward

28. When it comes to driving safely, what does it mean for drivers to communicate?

☐ **A)** Making use of hand signals

☐ **B)** Getting a helper before backing up a vehicle

☐ **C)** Making use of a cell phone or CB radio to talk to others

☐ **D)** Making use of the vehicle's 4-way flashers, turn signals, horn, and headlights

29. If you have to stop your vehicle to load passengers or cargo, you should...

☐ **A)** Flash the vehicle's brake lights to serve as a warning for drivers behind you

☐ **B)** Slow down the vehicle but do not bring it to a complete halt

☐ **C)** Swerve slowly from side to side to warn other drivers

☐ **D)** Brake suddenly to get the attention of drivers behind you

30. Where can the maximum weight loading for a set tire pressure be found?

☐ **A)** In the CDL manual

☐ **B)** On the side of each tire

☐ **C)** On the vehicle's dashboard

☐ **D)** On your manifest

31. While driving, you should always look up ahead for two main things. They are...

☐ **A)** Tunnels and overpasses

☐ **B)** Detours and law enforcement officers

☐ **C)** Gas stations and truck stops

☐ **D)** Traffic signs and other vehicles

32. What should be done as soon as the trailer starts going off the right path while backing?

☐ **A)** The top of the steering wheel should be turned in the drift's direction

☐ **B)** The top of the steering wheel should be turned in the opposite direction of the drift

☐ **C)** The steering wheel should be released and the vehicle's brakes engaged

☐ **D)** The steering wheel should be held tighter

33. When starting down a long downgrade, which of the factors below should not influence the driver's speed?

☐ **A)** eather and road conditions

☐ **B)** The driver's schedule

☐ **C)** How steep the downgrade is

☐ **D)** The vehicle's total weight including its cargo

34. During nighttime driving, you should...

☐ **A)** Always have the vehicle's high beams on

☐ **B)** Keep your windows rolled down to allow the fresh air keep you alert

☐ **C)** Have some coffee or any other caffeinated products so that you can remain awake

☐ **D)** Ensure your speed is slow enough to enable you stop within the headlights' range if there is an emergency

35. What should be done as a driver approaches a work crew on a highway?

☐ **A)** They should slow down

☐ **B)** They should sound the horn repeatedly

☐ **C)** They should rev the engine

☐ **D)** They should flash the vehicle lights

36. In a 55 mph zone where traffic is moving at 35 mph, what is an idea driving speed?

☐ **A)** 45 mph

☐ **B)** 50 mph

☐ **C)** 35 mph

☐ **D)** 55 mph

37. One of the statements below about a vehicle's spare tire is false

☐ **A)** It must have the same manufacturer as the main tires

☐ **B)** It must be the right size

☐ **C)** It must be free from any damages

☐ **D)** It must be correctly inflated

38. All of these are signs of a tire blowout apart from...

☐ **A)** A smoky odor

☐ **B)** A loud noise

☐ **C)** The vehicle vibrating

☐ **D)** A thumping noise

39. Which of the substances below can impair a driver's ability to drive safely?

☐ **A)** Certain prescription medicines

☐ **B)** Alcoholic drinks

☐ **C)** Certain cold medicines

☐ **D)** All of the above

40. Which of the statements below about a GPS device is true?

- ☐ **A)** Its screen should be more than 4 inches in size
- ☐ **B)** It should be equipped with both visual and audible warnings
- ☐ **C)** It should be manufactured in the USA
- ☐ **D)** It should be specially made for truck navigation

41. What should a driver do when approaching a railroad crossing with a heavy load?

- ☐ **A)** Speed up to cross the tracks more quickly.
- ☐ **B)** Stop, look both ways, and listen for a train before crossing.
- ☐ **C)** Keep the same speed and cross without stopping.
- ☐ **D)** Only stop if a train is visible.

42. Which of these is not a sign of driver fatigue?

- ☐ **A)** Frequent yawning.
- ☐ **B)** Increased focus and alertness.
- ☐ **C)** Trouble keeping your eyes open.
- ☐ **D)** Drifting from your lane.

43. For a commercial vehicle, what is the minimum tread depth for a steer tires?

- ☐ **A)** 2/32 of an inch.
- ☐ **B)** 4/32 of an inch.
- ☐ **C)** 1/32 of an inch.
- ☐ **D)** 3/32 of an inch.

2_ Pre-Trip Inspection Practice Test

This full-length CDL Pre-Trip Inspection practice test contains 47 questions. This test contains more CDL Pre-Trip Inspection practice questions than you'll probably need to pass the Pre-Trip Inspection endorsement, but why take any chances? Answering 47 practice questions will prepare you better than answering a mere 10 questions!

1. The engine compartment should not have more than...

- ☐ **A)** ½ inch play at the center of the belt
- ☐ **B)** ¾ inch play at the center of the belt
- ☐ **C)** ¾ incah play at the middle of the belt
- ☐ **D)** 2/4 inch play at the end of the belt

2. In order to check the power steering fluid, one should...

- ☐ **A)** Remove the oil fill cap and check the level
- ☐ **B)** Check the reservoir sight glass
- ☐ **C)** Check the power steering fluid dipstick
- ☐ **D)** All of the above

3. During the inspection of the dipstick to ensure that the lubrication level is above the refill mark, one is checking the...

- ☐ **A)** Power steering fluid level
- ☐ **B)** The oil level
- ☐ **C)** A and B
- ☐ **D)** The engine system

4. Puddles on the ground or dripping fluids on the underside of the engine and transmission indicates...

- ☐ **A)** A leaking hose
- ☐ **B)** A water pump problem
- ☐ **C)** An engine problem
- ☐ **D)** None of the above

5. In step three of the pre-trip inspection (starting of the engine and cab inspection), what is to be done first?

- ☐ **A)** Get into the car and start the engine
- ☐ **B)** Put gear shift into neutrality
- ☐ **C)** Get into the car and make sure the parking brakes are on
- ☐ **D)** None of the above

6. The normal oil temperature range of engine oil is...

- ☐ **A)** 20 - 35 psi
- ☐ **B)** 30 - 70 psi
- ☐ **C)** 10 - 20 psi
- ☐ **D)** 60 - 90 psi

7. Which part of the gauge has a governor cut?

- ☐ **A)** The air
- ☐ **B)** The oil pressure
- ☐ **C)** The voltmeter
- ☐ **D)** The nuts

8. Which one of these should be cleaned and adjusted from the inside?

- ☐ **A)** The mirrors
- ☐ **B)** The nuts
- ☐ **C)** The windshield
- ☐ **D)** None of the above

9. When the pedal is pumped thrice and held down for five seconds making sure that the pedal does not move, what is being inspected?

- ☐ **A)** The air brakes
- ☐ **B)** The hydraulic brake
- ☐ **C)** The parking brake
- ☐ **D)** None of the above

10. What belt should one make sure is securely mounted and adjusted and is not ripped off or frayed?

- ☐ **A)** The safety belt
- ☐ **B)** The air compressor
- ☐ **C)** The power steering
- ☐ **D)** None of the above7

11. ____ gauge should come up to normal within seconds after the engine is started

- ☐ **A)** Oil pressure
- ☐ **B)** Air pressure
- ☐ **C)** Coolant pressure
- ☐ **D)** All of the above

12. ____ gauge pressure should build from 50 - 90 psi within 3 minutes

- ☐ **A)** Air pressure
- ☐ **B)** Oil pressure
- ☐ **C)** Coolant pressure
- ☐ **D)** None of the above

13. When checking for damage to the longitudinal numbers, cross members, what are you inspecting?

- ☐ **A)** Frame
- ☐ **B)** Fuel tank
- ☐ **C)** Exhaust system
- ☐ **D)** None of the above

14. What transforms steering column action into wheel action?

- ☐ **A)** Steering box and hoses
- ☐ **B)** Suspension
- ☐ **C)** Steering link
- ☐ **D)** None of the above

15. What is the best way to check the coolant level?

- ☐ **A)** Check the sight glass
- ☐ **B)** Check the dipstick
- ☐ **C)** Check the temperature gauge
- ☐ **D)** All of the above

16. Make sure the ______ are not cut, chaffed, or worn out and not dragged against tractor parts

- ☐ **A)** Locking jaws
- ☐ **B)** Door ties and lift
- ☐ **C)** Air lines and electrical lines
- ☐ **D)** All of the above

17. One of these keeps your vehicle from rolling when parked

- ☐ **A)** The parking brake
- ☐ **B)** The locking pin
- ☐ **C)** The service brake
- ☐ **D)** The gear system

18. What carries air to the wheel brake assembly?

- ☐ **A)** Slack adjusters
- ☐ **B)** Brake hoses and lines
- ☐ **C)** Brake chambers
- ☐ **D)** None of the above

19. One of these converts air pressure to mechanical force to operate the brake at the wheels

- ☐ **A)** Brake chambers
- ☐ **B)** Drum brake
- ☐ **C)** Brake hoses
- ☐ **D)** All of the above

20. A driver can easily check that the vehicle's oil level is adequate if there is a sight glass on the wheel. Otherwise, they should look for leaks around the...

☐ **A)** Battery

☐ **B)** Hub oil seals and axle seals

☐ **C)** Hydraulic brakes

☐ **D)** All of the above

21 Which external lights should one check?

☐ **A)** Brake lights

☐ **B)** Turn signals

☐ **C)** All of the above

☐ **D)** None of the above

22. What is responsible for the transmission to the drive axle?

☐ **A)** The steering play

☐ **B)** The driveshaft

☐ **C)** The air compressor belt

☐ **D)** None of the above

23. What is responsible for the heating of the cab and prevents frosting from forming on the windshield?

☐ **A)** The header board

☐ **B)** The heater / the defroster

☐ **C)** The air compressor

☐ **D)** All of the above

24. What do you inspect to make sure all are present and show no signs of looseness such as rust trails?

- ☐ **A)** Rims
- ☐ **B)** Lug nuts
- ☐ **C)** Dodger
- ☐ **D)** None of the above

25. What should one make sure is tightly secured and the taps are tight?

- ☐ **A)** Air compressor
- ☐ **B)** Fuel tanks
- ☐ **C)** Water pump
- ☐ **D)** The steering gear

26. What structure supporting the fifth wheel skid plate should be checked for cracks and breaks?

- ☐ **A)** Platform
- ☐ **B)** Kingpin
- ☐ **C)** Apron
- ☐ **D)** The gear shift

27. Which of the following can disengage the engine from the drive train?

- ☐ **A)** The driveshaft
- ☐ **B)** The parking brake
- ☐ **C)** The gearshift and clutch
- ☐ **D)** The gear pin

28. Which gauge helps you prevent engine failure, seizure, or breakdown?

- ☐ **A)** The oil pressure
- ☐ **B)** The ammeter
- ☐ **C)** The power steering
- ☐ **D)** None of the above

29. Which of the following is not part of inspecting a vehicle's coupling?

- ☐ **A)** Checking that the fifth wheel jaw has closed on the correct part of the king pin
- ☐ **B)** Driving a few feet to test if it is connected correctly
- ☐ **C)** All of the above
- ☐ **D)** None of the above

30. What is responsible for maintaining air pressure in the air brake system?

- ☐ **A)** The alternator
- ☐ **B)** The water pump
- ☐ **C)** The air compressor
- ☐ **D)** All of the above

31. If you cannot make a right turn without swinging into another lane, what should be done?

- ☐ **A)** Turn wide as you complete the turn
- ☐ **B)** Turn wide before starting the turn
- ☐ **C)** Find some other location to turn
- ☐ **D)** None of the above

32. Which of the following is not true about an Antilock Braking System?

☐ **A)** It only activates when the wheels are in danger of locking up

☐ **B)** It increases the vehicle's normal braking ability

☐ **C)** It is found on all tractors manufactured after the year 1998

☐ **D)** All of the above

33. What helps in releasing the fifth wheel locking jaw so that the trailer can be coupled?

☐ **A)** The release arm

☐ **B)** The locking pin

☐ **C)** The draw bar

☐ **D)** The brake system

34. What is the sliding mechanism for tandem axles on trailers?

☐ **A)** The locking pins

☐ **B)** The landing gear

☐ **C)** The suspension system

☐ **D)** All of the above

35. Which of a vehicle's tires should have a minimum tread depth of 4/32 inches?

☐ **A)** The spare tires

☐ **B)** The non-drive axle tires

☐ **C)** The steering axle tires

☐ **D)** None of the above

36. What should be clean with no dirt, illegal sticker, or any form of obstruction?

- ☐ **A)** The headlights
- ☐ **B)** The windshield and mirrors
- ☐ **C)** The light indicators
- ☐ **D)** All of the above

37. What is to be checked that it is free from damages and securely bolted to the tractor frame?

- ☐ **A)** The catwalk
- ☐ **B)** The kingpin
- ☐ **C)** The apron
- ☐ **D)** None of the above

38. What is responsible for the carrying of hydraulic fluid to the wheel brake assembly?

- ☐ **A)** The slack adjuster
- ☐ **B)** The brake linings
- ☐ **C)** The brake chambers
- ☐ **D)** The brake hoses and lines

39. _____ are the brackets, bolts, or bushings that are used to attach the spring to the axle or vehicle frame

- ☐ **A)** The torque
- ☐ **B)** The spring mounts
- ☐ **C)** The mounting bolts
- ☐ **D)** The tongue

40. What potential issues should a driver be aware of when coupling a trailer that is positioned too high?

☐ **A)** The fifth wheel might slide under the trailer, failing to couple correctly.

☐ **B)** The nose of the trailer could be damaged during the coupling process.

☐ **C)** The undercarriage of the tractor could be struck and damaged.

☐ **D)** All of the above are potential issues.

41. When doing the pre-trip inspection, you should tell the examiner each item you are checking and two or three reasons why you're checking them. True or false?

☐ **A)** True

☐ **B)** False

☐ **C)** Maybe

☐ **D)** Only if asked

42. Should an overview of your vehicle be done as you approach it (including looking for leaks, leaning, or excessive damage)?

☐ **A)** Yes

☐ **B)** No

☐ **C)** Maybe

☐ **D)** Only if the vehicle had mechanical issues during the last trip

43. When opening the hood, you should check...

☐ **A)** The hinges, catches, and springs

☐ **B)** Nothing

☐ **C)** The windshield

☐ **D)** Both A and C

44. When checking the mirrors on the outside of the vehicle, you should ensure that...

- ☐ **A)** They are properly mounted
- ☐ **B)** There are no cracks
- ☐ **C)** They are clean
- ☐ **D)** All the above

45. What should you check regarding the air pressure gauge during a pre-trip inspection?

- ☐ **A)** The gauge should register zero.
- ☐ **B)** The gauge should show increasing pressure until the system reaches its operating range.
- ☐ **C)** The gauge should remain constant at all times.
- ☐ **D)** The gauge should not be checked during a pre-trip inspection.

46. During a pre-trip inspection, how should you verify the condition of the leaf springs?

- ☐ **A)** Ensure they are painted.
- ☐ **B)** Check for missing or shifted leaves, and look for signs of wear or damage.
- ☐ **C)** Make sure they are covered.
- ☐ **D)** Verify that they are lubricated.

47. What is important to inspect on the exhaust system during a pre-trip inspection?

- ☐ **A)** The color of the exhaust fumes.
- ☐ **B)** The presence of a muffler.
- ☐ **C)** No loose, broken, or missing parts and no signs of leaks.
- ☐ **D)** The sound it makes when the engine is running.

3_ Air Brakes Practice Test

This full-length CDL Air Brakes practice test contains 47 questions. This test contains more CDL Air Brakes practice questions than you'll probably need to pass the Air Brakes endorsement, but why take any chances? Answering 47 practice questions will prepare you better than answering a mere 10 questions!

1. What kind of air is needed in air brakes?

- ☐ **A)** Heated air
- ☐ **B)** Decomposed air
- ☐ **C)** Warm air
- ☐ **D)** Compressed air

2. What 3 types of brake systems are used in commercial vehicles' brake systems?

- ☐ **A)** The service brake, the emergency brake, and the parking brake
- ☐ **B)** The anti-lock brake, the service brake, and the emergency brake
- ☐ **C)** The hand brake, parking brake, and the emergency brake
- ☐ **D)** The service brake, the hand brake, and the emergency brake

3. What pumps air into the air storage tank?

- ☐ **A)** The safety valve
- ☐ **B)** The air compressor
- ☐ **C)** The governor
- ☐ **D)** The evaporator

4. One of the following statements is true about spring brakes

- ☐ **A)** It will not function if the system loses air pressure
- ☐ **B)** They are held by mechanical force
- ☐ **C)** They come on at 60 psi
- ☐ **D)** They make use of air pressure to stop vehicles

5. At what psi level is the safety relief valve supposed to be set to be safe to open at?

- ☐ **A)** 125 psi
- ☐ **B)** 60 psi
- ☐ **C)** 150 psi
- ☐ **D)** 100 psi

6. One of the following is the most common type of foundation brake

- ☐ **A)** Disc brake
- ☐ **B)** Wedge brake
- ☐ **C)** The anti-lock brake
- ☐ **D)** S-Cam drum brake

7. What is the basic function of the alcohol evaporator?

- ☐ **A)** It is for the purpose of heating the air tank
- ☐ **B)** It is to put alcohol into the air system
- ☐ **C)** It is to remove water and oil from the brake system
- ☐ **D)** It is to keep alcohol out of the vehicle parts

8. One of the following is true about the front wheel braking

- ☐ **A)** The front wheel braking is good under all conditions
- ☐ **B)** It is ineffective under slippery conditions
- ☐ **C)** Front wheel braking will cause skids on ice
- ☐ **D)** One should put a front brake limiting valve if equipped on the slippery positions in adverse conditions

9. What is the perfect time to drain one's air tanks?

- ☐ **A)** The end of a trip
- ☐ **B)** The end of the month
- ☐ **C)** The end of every working day
- ☐ **D)** The end of the fiscal quarter

10. On what kind of vehicle is a low-pressure warning signal required to be?

- ☐ **A)** Buses only
- ☐ **B)** Vehicles with air brakes
- ☐ **C)** Straight trucks
- ☐ **D)** The gear

11. What causes a vehicle's brake to fail?

- ☐ **A)** The brakes being out of adjustment
- ☐ **B)** The excessive use of service brakes
- ☐ **C)** Not relying enough on the engine brake
- ☐ **D)** All of the above

12. Modern air brake systems are made up of three different systems: the service brakes, parking brakes, and _______ brakes.

- ☐ **A)** S-Cam
- ☐ **B)** Emergency
- ☐ **C)** Foot
- ☐ **D)** All of the above

13. What is the maximum acceptable leakage rate per minute after the initial drop?

- ☐ **A)** 5 psi for single vehicles and 6 psi for combination vehicles respectively
- ☐ **B)** 1 psi for single vehicles and 7 psi for combination vehicles respectively
- ☐ **C)** 3 psi for single vehicles and 4 psi for combination vehicles respectively
- ☐ **D)** 5 psi for single vehicles and 10 psi for combination vehicles respectively

14. How are the truck slack adjusters checked?

- ☐ **A)** Press the brake pedal while listening for any strange noise
- ☐ **B)** Pull hard on each slack adjuster you can reach
- ☐ **C)** Accelerate and then brake hard
- ☐ **D)** Do all of the above

15 In air brake vehicles, parking brakes should be used...

- ☐ **A)** Whenever the vehicle is left unattended
- ☐ **B)** As little as possible
- ☐ **C)** During pre-trip and post-trip inspection
- ☐ **D)** None of the above

16. What is the function of a supply pressure gauge?

- ☐ **A)** To show how hot the air in the tank is
- ☐ **B)** To show how much air is in the tank
- ☐ **C)** To warn if the air in the tank is too low
- ☐ **D)** All of the above

17. At what measurement will the safety valve open?

- ☐ **A)** 100 psi
- ☐ **B)** 250 psi
- ☐ **C)** 200 psi
- ☐ **D)** 150 psi

18. In what manner do brakes work especially on a long steep downgrade?

- ☐ **A)** They work on the main braking system
- ☐ **B)** They work as a supplement to the braking of the engine
- ☐ **C)** They work as the main braking mechanism with the engine braking effect as an emergency backup
- ☐ **D)** They are not applicable

19. One of the following is not a part of the drum brake

- ☐ **A)** The return brake
- ☐ **B)** The safety valve
- ☐ **C)** The slack adjusters
- ☐ **D)** The brake drum

20. What will happen if the air tanks are not drained?

- ☐ **A)** One will drive too quickly
- ☐ **B)** Brakes may fail cause of water freezing
- ☐ **C)** Transmission fluid may drain out
- ☐ **D)** The vehicle's left side brake will cease to operate

21. Vehicles with air brakes have...

- ☐ **A)** A supply pressure gauge
- ☐ **B)** An air use gauge
- ☐ **C)** Backup hydraulic system
- ☐ **D)** None of the above

22. How should the air leakage rate be tested?

- ☐ **A)** Turn the engine off and release of the parking brake
- ☐ **B)** Charge the air system completely and leave the engine running
- ☐ **C)** Leave the engine running and release the parking brake
- ☐ **D)** Depress the brake pedal while keeping an eye on the air pressure gauge

23. What should be done before leaving one's vehicle?

- ☐ **A)** Chock the wheels
- ☐ **B)** Remove the keys
- ☐ **C)** Set the parking brakes
- ☐ **D)** Do all of the above

24. One of the following is not a part of the air brake system

- ☐ **A)** The emergency air brake system
- ☐ **B)** The radio signal
- ☐ **C)** The parking brake system
- ☐ **D)** The emergency brake system

25. When is the parking brake required to be used?

- ☐ **A)** Every time you leave your vehicle with few exceptions
- ☐ **B)** Only if you are away from your vehicle for a selected period of time
- ☐ **C)** Only in urban areas
- ☐ **D)** Only when there are other motorists around

26. What set of vehicles should have low air pressure signals?

- ☐ **A)** Vehicles built after the year 2010
- ☐ **B)** Vehicles built after the year 2015
- ☐ **C)** All vehicles with air brakes that are currently in operation
- ☐ **D)** None

27. One of the following is true about a dual air brake system

- ☐ **A)** It uses a single set of brake controls
- ☐ **B)** One of the systems is called "primary" while the other is called "secondary"
- ☐ **C)** One system operates the front axle and the other operates the rear axle
- ☐ **D)** All of the above

28. In what conditions are the front wheel brakes good?

- ☐ **A)** In good weather
- ☐ **B)** In every weather condition
- ☐ **C)** Icy conditions only
- ☐ **D)** None of the above

29. How best can one check if their vehicle's spring brakes come on automatically?

- ☐ **A)** Step on and off the brake pedal until the manufacturer's low psi specification is met, in order for springs to deploy
- ☐ **B)** Step on and off the brake pedal until the parking valve pops out
- ☐ **C)** On single vehicles, continue to step on and off the brake pedal until the parking valve pops out
- ☐ **D)** All of the above

30. What is an Anti-lock braking system?

- ☐ **A)** It activates when wheels are about to lock up
- ☐ **B)** It increases one's normal braking capability
- ☐ **C)** It shortens one's stopping distance
- ☐ **D)** It decreases normal braking capability

31. One of the following is true about the brake function when the anti-lock brake system fails

- ☐ **A)** It will slow the truck to a halt and cause you to pull over
- ☐ **B)** There will be a lack of brake function and the truck will be out of control
- ☐ **C)** You will still be able to use your brakes normally, but you will need to have the ABS serviced as soon as possible
- ☐ **D)** It may cause a huge fire risk

32. Using brakes on a downgrade acts as a supplement to...

- ☐ **A)** The use of the spring brakes
- ☐ **B)** The braking effect of the engine
- ☐ **C)** The use of the front brake limiting valve
- ☐ **D)** All of the above

33. In a vehicle that has a properly functioning dual air brake system and minimum-sized air tanks, the air pressure would build from 85 - 100 psi within how many seconds?

- ☐ **A)** 45
- ☐ **B)** 60
- ☐ **C)** 20
- ☐ **D)** 70

34. What will happen if the air pressure warning signal is not working?

- ☐ **A)** There will be a sudden emergency braking in the single circuit air system
- ☐ **B)** Air pressure will be lost
- ☐ **C)** Nothing will happen
- ☐ **D)** There will be a skid

35. The braking power of the spring brake...

- ☐ **A)** Increases when the service brakes are hot
- ☐ **B)** Depends on whether the service brakes are in proper adjustment
- ☐ **C)** Is unaffected by the condition of the service brakes
- ☐ **D)** Depends on the gas

36. You would know if your brakes are fading off if...

- ☐ **A)** You release the brake pedal and the vehicle's speed increases
- ☐ **B)** You have to press the brake pedal harder than usual to control your speed
- ☐ **C)** The brake feels spongy when you apply pressure
- ☐ **D)** All of the above

37. What does the service line in a vehicle well-equipped with air brakes do?

☐ **A)** It carries fuel at the filling station

☐ **B)** It supplies air to the trailer air tanks

☐ **C)** It carries air from the vehicle's foot brake or the trailer hand brake

☐ **D)** None of the above

38. In perfect conditions, the average driver of a vehicle who is equipped with air brakes and traveling at 55mph would need to stop at what distance?

☐ **A)** More than 300 feet

☐ **B)** 150 -250 feet

☐ **C)** 50 - 150 feet.

☐ **D)** All of the above

39. The application of the pressure gauge shows the amount of air pressure...

☐ **A)** In the tanks

☐ **B)** In the modulating control valve

☐ **C)** Being applied to the brakes

☐ **D)** Nothing happens

40. For what reason should one not fan the brake when driving on a long downgrade?

☐ **A)** The brakes won't cool

☐ **B)** Air usage becomes less when fanning

☐ **C)** Brake linings do not get hot when fanning

☐ **D)** None of the above

41. Where is the safety valve situated in a vehicle well-equipped with air brakes?

- ☐ **A)** The dashboard
- ☐ **B)** The first tank from which the air compressor pumps out
- ☐ **C)** Under the brake pedal
- ☐ **D)** In the second tank from which the air compressor pumps out

42. In a walk-around inspection, check the brake drums to be sure that no cracks are longer than ______ width of the friction area

- ☐ **A)** One-half
- ☐ **B)** One-third
- ☐ **C)** One-quarter
- ☐ **D)** All of the above

43. In a vehicle that is well equipped with air brakes, what activates the stoplight switch when you brake?

- ☐ **A)** The air pressure
- ☐ **B)** The spring pressure
- ☐ **C)** The mechanical force
- ☐ **D)** None of the above

44. The brake pads should be _____ for the brakes to be on

- ☐ **A)** Worn 1/32 of an inch
- ☐ **B)** Against the drum
- ☐ **C)** Disconnected from the slack adjusters
- ☐ **D)** Worn dangerously thin

45. What is the primary function of the air compressor in an air brake system?

☐ **A)** To release air from the tanks.

☐ **B)** To maintain air pressure in the air tanks.

☐ **C)** To cool the air in the system.

☐ **D)** To lubricate the brake mechanisms.

46. What is the purpose of the air dryer in an air brake system?

☐ **A)** To dry the fuel.

☐ **B)** To clean the fuel in the vehicle's engine.

☐ **C)** To remove moisture from the air before it enters the air tanks.

☐ **D)** To regulate air temperature.

47. Why is it important to regularly drain water from the air tanks in an air brake system?

☐ **A)** To prevent water from freezing in the tanks and cause brake failure.

☐ **B)** To reduce air pressure.

☐ **C)** To improve fuel efficiency.

☐ **D)** To clean the tanks.

4_ Tanker Vehicles Practice Test

This full-length CDL Tanker Vehicles practice test contains 45 questions. This test contains more CDL Tanker Vehicles practice questions than you'll probably need to pass the Tanker Vehicles endorsement, but why take any chances? Answering 45 practice questions will prepare you better than answering a mere 10 questions!

1. One of the following is true about managing spaces from side to side

- ☐ **A)** Avoid traveling alongside other vehicles
- ☐ **B)** Keep your vehicle to the right side of your lane
- ☐ **C)** Strong winds make it easier to stay in one's lane
- ☐ **D)** All of the above

2. When is a tank endorsement required?

- ☐ **A)** They are required for vehicles that transport liquid and gases
- ☐ **B)** They are required for vehicles that transport passengers
- ☐ **C)** They are required for hazardous materials transportation
- ☐ **D)** They are required for air system

3. In what way does a liquid surge affect the handling of a tanker?

- ☐ **A)** It increases the wind drag of the tank
- ☐ **B)** The surge allows one to turn corners tighter
- ☐ **C)** It can move the truck in the direction of the liquid waves move
- ☐ **D)** All of the above

4. When loading small tanks of a cargo tank equipped with bulkheads, one should check for...

- ☐ **A)** The distribution of weight
- ☐ **B)** The water content
- ☐ **C)** The air-to-fuel ratio
- ☐ **D)** None of the above

5. One should have the knowledge of the outage needed for the liquids carried because...

- ☐ **A)** The heaviest liquids do not need any outage
- ☐ **B)** Tank baffles are not always legal with outage
- ☐ **C)** Some liquids expand more when they are warm
- ☐ **D)** All of the above

6. The best way to take a curve in a tanker is to slow to a safe speed before entering the curve and then ______

- ☐ **A)** Brake lightly
- ☐ **B)** Speed up slightly
- ☐ **C)** Downshift twice
- ☐ **D)** Speed up high

7. How often should one check the tires of a placarded vehicle with dual tires?

- ☐ **A)** Every 5 hours
- ☐ **B)** At the beginning of each trip and each time one stops
- ☐ **C)** Every week
- ☐ **D)** Every 1 hour

8. One should be very cautious when driving smoothbore tankers, especially when one is...

- ☐ **A)** Driving against the wind
- ☐ **B)** Starting or stopping
- ☐ **C)** Driving uphill or downhill
- ☐ **D)** Sliding

9. A vehicle carrying hazardous materials must have a placard on how many sides?

- ☐ **A)** 4
- ☐ **B)** 3
- ☐ **C)** 2
- ☐ **D)** 7

10. A smoothbore hole tank is also referred to as...

- ☐ **A)** Aluminium tank
- ☐ **B)** Unbaffled tank
- ☐ **C)** Baffled tank
- ☐ **D)** All of the above

11. Tankers that haul liquid require special care due to what reasons?

- ☐ **A)** The liquid movement
- ☐ **B)** The high center of gravity
- ☐ **C)** All of the above
- ☐ **D)** False, there is no need to be extra careful when transporting liquid cargo

12. When driving on a clear night, a driver must dim their headlights from high to low. They should also adjust their speed so they can stop within...

- ☐ **A)** The distance they can see ahead
- ☐ **B)** The length of the vehicle
- ☐ **C)** The distance to be covered in the next 15 seconds
- ☐ **D)** None of the above

13. One of these is true about tank vehicles

- ☐ **A)** The posted speed for a curve will be too fast for a tank vehicle
- ☐ **B)** The emergency brakes in a tank vehicle can be used to avoid crashes
- ☐ **C)** Loaded tanks take longer to stop than empty ones
- ☐ **D)** All of the above

14. The person who is qualified and is watching the offloading and loading of the tank should be...

- ☐ **A)** Within 30 feet of the tank
- ☐ **B)** Within 25 feet of the tank
- ☐ **C)** Within 35 feet of the tank
- ☐ **D)** None of the above

15. What does an outage mean?

- ☐ **A)** The allowance for the expansion of liquid
- ☐ **B)** The liquid weight
- ☐ **C)** How fast the liquid tanks drain
- ☐ **D)** All of the above

16. What do you do to steer your tanker quickly in order to avoid a hazard?

- ☐ **A)** You should place the gear in "Neutral"
- ☐ **B)** You should avoid braking as you turn
- ☐ **C)** You should brake hard and fast
- ☐ **D)** None of the above

17. What is a baffled tanker?

- ☐ **A)** They are separate tanks inside a tanker
- ☐ **B)** They are hollow balls floating in the liquid to slow movement
- ☐ **C)** They are bulkheads with holes that allow the liquid to flow through
- ☐ **D)** All of the above

18. What does the amount of liquid allowed to be loaded in a tank depend on?

- ☐ **A)** The amount of liquid it would expand to in transit
- ☐ **B)** The legal weight limit
- ☐ **C)** All of the above
- ☐ **D)** None of the above

19. One of the statements below about tankers is true

☐ **A)** They can turn over at the speed limit posted for curves

☐ **B)** The load weight is carried high up off the road

☐ **C)** All are true

☐ **D)** None of the above

20. One of these is a true statement about stopping distance and speed

☐ **A)** One needs about twice as much stopping distance at 40mph than at 20mph

☐ **B)** Wet roads double the stopping distance at any speed

☐ **C)** Both options are true

☐ **D)** All of the above

21. A side-to-side surge can cause:

☐ **A)** Over speeding

☐ **B)** A rollover

☐ **C)** Suspension system failure

☐ **D)** None of the above

22. When an emergency forces you to stop your tanker quickly to avoid a crash, what should you do?

☐ **A)** Make use of the emergency brakes

☐ **B)** Make use of controlled or stab braking

☐ **C)** Brake hard until you lock the wheels

☐ **D)** All of the above

23. The effects of a liquid surge is reduced by...

☐ **A)** The hauling of thicker liquids

☐ **B)** Starting, turning, and stopping the vehicle smoothly

☐ **C)** All of the above

☐ **D)** None of the above

24. Which of the following makes a liquid tanker harder to handle?

- ☐ **A)** The width of the cargo
- ☐ **B)** The high center of the gravity of the cargo
- ☐ **C)** The low center of the gravity of the cargo
- ☐ **D)** All of the above

25. A smoothbore tanker is used in transporting:

- ☐ **A)** Heavy liquid
- ☐ **B)** Liquid or milk products
- ☐ **C)** Acidic liquid
- ☐ **D)** None of the above

26. When should the escape ramp be used by a driver in the occurrence of him losing his brakes?

- ☐ **A)** Never
- ☐ **B)** Always
- ☐ **C)** If the tank is baffled
- ☐ **D)** Only if there are other vehicles around

27. If a vehicle is placarded, one must stop how many feet before the nearest rail at a railroad crossing?

- ☐ **A)** 15 - 50
- ☐ **B)** 10 - 36
- ☐ **C)** 5 - 20
- ☐ **D)** 15 - 65

28. What kind of surge can liquid in a tank with baffles have?

- ☐ **A)** Front-to-back
- ☐ **B)** Side-to-side
- ☐ **C)** Top-to-bottom
- ☐ **D)** None of the above

29. How do you expect a truck with a cargo tank that has baffles to handle on the road?

- ☐ **A)** To have lesser forward and backward surge than a tank without baffles
- ☐ **B)** The truck will seem heavier than it really is
- ☐ **C)** It will handle the same way as a tanker without baffles
- ☐ **D)** None of the above

30. Tanker drivers require special skills because...

- ☐ **A)** Liquid surge occurs in tankers
- ☐ **B)** Tankers have an increased risk of rollover
- ☐ **C)** All the above reasons
- ☐ **D)** False, tanker drivers do not need to have any special skills

31. Which of the following is a potential danger of a side-to-side surge?

- ☐ **A)** The suspension of the system failure
- ☐ **B)** Rollover
- ☐ **C)** Contentment breach
- ☐ **D)** None of the above

32. How should a tanker truck be driven in a curved place?

- ☐ **A)** One should accelerate before the curve and then downshift as you go through the curve
- ☐ **B)** One should use controlled braking as you go through the curve
- ☐ **C)** One should slow down before the curve and then accelerate as you go through the curve
- ☐ **D)** None of the above

33. One of the following is a divider between compartments inside of a tank

- ☐ **A)** Separation walls
- ☐ **B)** Baffles
- ☐ **C)** Bulkhead
- ☐ **D)** None of the above

34. Dividers inside tankers have holes that are in them and are called...

☐ **A)** Barriers

☐ **B)** Baffles

☐ **C)** Separators

☐ **D)** None of the above

35. Manhole covers should be...

☐ **A)** Left partially open to prevent a buildup of fumes

☐ **B)** Opened and reclosed before driving the vehicle

☐ **C)** Closed before driving the vehicle

☐ **D)** All of the above

36. How exactly should a tank vehicle be inspected?

☐ **A)** Do it the same way you inspected other tank vehicles before

☐ **B)** Read the vehicle's manual to learn how to do it

☐ **C)** Ask a fellow driver how to do it

☐ **D)** All of the above

37. Liquid surge is especially dangerous...

☐ **A)** On narrow city streets

☐ **B)** On hills

☐ **C)** On slippery roads

☐ **D)** All of the above

38. Drivers hauling liquids in tankers will need to be extra careful for two main reasons. One of them is the _____ center of gravity that tankers have.

☐ **A)** Flat

☐ **B)** Wide

☐ **C)** High

☐ **D)** None of the above

39. You must have a tank vehicle (N) endorsement to drive a vehicle with...

☐ **A)** Individual tanks over 225 gallons or a total of at least 2,400 gallons

☐ **B)** Individual tanks over 119 gallons or a total of at least 1,000 gallons

☐ **C)** Individual tanks over 65 gallons or a total of at least 500 gallons

☐ **D)** All of the above

40. Which of these is the most important thing to remember about emergency braking?

☐ **A)** If the wheels are skidding, you cannot control the vehicle

☐ **B)** Disconnecting the steering axle brakes will help keep your vehicle in a straight line

☐ **C)** Never engage a vehicle's emergency brakes without downshifting first

☐ **D)** All of the above

41. Which of these equipment needs to function properly if they are present in your vehicle?

☐ **A)** Vapor recovery kit

☐ **B)** Grounding and bonding cables

☐ **C)** Emergency shut-off systems

☐ **D)** All of the above

42. If a hill or curve prevents drivers behind you from seeing the vehicle within 500 feet...

☐ **A)** You should move the rear reflective triangle further down the road to give early adequate warning to other motorists

☐ **B)** You do not need to put out reflective triangles unless If tthe vehicle will be stopped for less than 30 minutes, there is no need to put out reflective triangles

☐ **C)** The vehicles taillights should be kept on to serve as a warning for other drivers

☐ **D)** None of the above

43. What is a surge in a liquid tanker?

- ☐ **A)** Sudden acceleration of the tanker.
- ☐ **B)** Movement of liquid in partially filled tanks, affecting the vehicle's stability.
- ☐ **C)** Braking action in tankers.
- ☐ **D)** A weather phenomenon.

44. What is the most important factor to consider when loading a liquid tanker?

- ☐ **A)** The color of the liquid.
- ☐ **B)** The weight distribution of the liquid.
- ☐ **C)** The destination.
- ☐ **D)** The temperature of the liquid.

45. What is the purpose of baffles in a liquid tanker?

- ☐ **A)** To decorate the inside of the tank.
- ☐ **B)** To contain spills in case of an accident.
- ☐ **C)** To reduce liquid movement during transit.
- ☐ **D)** To increase the capacity of the tank.

Dear Valued Reader,

Congratulations on acquiring our CDL Practice Workbook! You've taken an important step towards mastering the material for your CDL exam. To enhance your learning journey, we're delighted to offer you exclusive access to additional resources on our e-learning platform, "LearniK."

As a valued owner of our workbook, here's what you can access:

» **Audiobook Version:** Complement your reading with the Audiobook version of the workbook. It's perfect for absorbing material during commutes or while multitasking, ensuring you make the most of your time.

» **Interactive Flashcards:** Boost your memory and retention with our interactive flashcards. These digital cards are a quick and efficient way to review key concepts, making them a great tool for your study sessions.

» **Engaging Online Videos:** Enhance your practical skills with our online videos, focusing on key aspects like the pre-trip inspection and other essential practices for the CDL practical exam. These videos provide concise, step-by-step demonstrations to help you master the hands-on skills needed for your test and on-the-road success.

To unlock these exclusive resources, simply scan the QR code provided in this workbook. This will grant you free access to the LearniK platform, where these additional study aids await you.

Remember, your success in the CDL exam is our top priority. By utilizing these tools in conjunction with our practice workbook, you're setting yourself up for a well-rounded and effective study experience.

Get started today! **Scan the QR code**, explore the wealth of resources on LearniK, and take your exam preparation to the next level.

5 _ Hazardous Materials Practice Test

This full-length CDL Hazardous Materials practice test contains 51 questions. This test contains more CDL Hazardous Materials practice questions than you'll probably need to pass the Hazardous Materials endorsement, but why take any chances? Answering 51 practice questions will prepare you better than answering a mere 10 questions!

1. When carrying a 100-pound cargo of silver cyanide, what precautions should be taken when given 100 cartons of battery acid to carry at a dock?

- ☐ **A)** Do not load the battery acid
- ☐ **B)** Make sure the silver cyanide is loaded on top of the battery acid
- ☐ **C)** Make sure the battery acid is loaded on top of the silver cyanide
- ☐ **D)** None of the above

2. For the hazard classes in Placard table 2, one should use them only if they are transporting a total of...

- ☐ **A)** 501 pounds and more
- ☐ **B)** 1,001 pounds and more
- ☐ **C)** 1,501 pounds and more
- ☐ **D)** None of the above

3. A technical name is...

- ☐ **A)** The name of a hazardous material used in scientific texts, and it is recognized as its biological name
- ☐ **B)** The medical name for hazardous materials used by medical personnel
- ☐ **C)** The name of a hazardous material used in the truck community
- ☐ **D)** None of the above

4. Placards should be used how many inches away from other markings?

☐ **A)** 3 inches

☐ **B)** 7 inches

☐ **C)** 12 inches

☐ **D)** 20 inches

5. What is necessary for the qualification of non bulk packaging?

☐ **A)** A maximum net mass of 400kg or less

☐ **B)** A maximum capacity of 450l or less for a liquid

☐ **C)** Both options above

☐ **D)** None of the above

6. When a truck carrying explosives has collided with another vehicle you should not pull them apart until...

☐ **A)** You have placed the explosives 200 feet away from the building in the environment

☐ **B)** 30 minutes have passed

☐ **C)** The shipper's foreman is present

☐ **D)** None of the above

7. A vehicle containing 500 pounds each of division 1.1 explosives and division 1.2 explosives must have...

☐ **A)** Blasting agent placards

☐ **B)** Dangerous substances placards

☐ **C)** Explosives placards

☐ **D)** All of the above

8. When can you be allowed to move an improperly placarded vehicle?

☐ **A)** Never

☐ **B)** During an emergency only

☐ **C)** If the load isn't truly hazardous

☐ **D)** Only when the cargo is very important

9. Hazard shipping papers should include the description of the hazardous material and...

☐ **A)** Monetary value

☐ **B)** Contact information of the driver's next of kin

☐ **C)** Emergency response information

☐ **D)** None of the above

10. Other places where the hazardous material identification must appear are...

☐ **A)** On all bulk packaging and the cargo tank

☐ **B)** On the gas tank and the sticker in the glove compartment

☐ **C)** On the temporary license holder and the steering wheel

☐ **D)** None of the above

11. Placarded vehicles should carry a fire extinguisher with a minimum rating of...

☐ **A)** 20 B:C

☐ **B)** 10 B:C

☐ **C)** 5 B:C

☐ **D)** None of the above

12. Smoking and performing of any similar activity are not allowed within 25 feet of...

☐ **A)** Class 4.2 only

☐ **B)** Class 5.2 only

☐ **C)** Class 1,2,3 and 4

☐ **D)** All of the above

13. On receiving a leaking package, the driver should...

☐ **A)** Refuse the package

☐ **B)** Transport it as it is

☐ **C)** Fix the package before transportation

☐ **D)** Hide it

14. In the determination of the use of placards, which of the following is not needed?

- ☐ **A)** The amount of the substance shipped
- ☐ **B)** The manufacturing date of the substance
- ☐ **C)** Amount of all the hazardous material of all classes being shipped
- ☐ **D)** None of the above

15. What is the function of the Emergency Response Guidebook?

- ☐ **A)** It is created by the department of transportation and is used nation-wide
- ☐ **B)** It is studied by emergency personnel to keep the public safe
- ☐ **C)** All of the above
- ☐ **D)** None of the above

16. One of the following is not an acceptable way of marking for hazardous materials

- ☐ **A)** Written in italics
- ☐ **B)** UN marks
- ☐ **C)** Written in Roman description
- ☐ **D)** Written boldly

17. Engines run a pump when you are delivering compressed gas. After delivery, when should the engine be turned off?

- ☐ **A)** Before unhooking
- ☐ **B)** After unhooking
- ☐ **C)** Turn it off on arrival: use the power to run the pump
- ☐ **D)** None of the above

18. If an "X" or "RQ" is in the "HM" column of the shipping paper entry, what does this imply?

- ☐ **A)** The material is regulated by hazardous material regulations
- ☐ **B)** The entry only refers to materials that must be top-loaded
- ☐ **C)** The material listed on the line is the largest part of the shipment
- ☐ **D)** All of the above

19. One of the following materials is an acceptable floor liner used in moving division 1.1 and 1.2 materials

- ☐ **A)** Stainless steel
- ☐ **B)** Carbon steel
- ☐ **C)** Non-ferrous metal
- ☐ **D)** None of the above

20. The major difference between a cargo tank and a portable tank is...

- ☐ **A)** Temporary versus permanent attachment
- ☐ **B)** Being filled on the vehicle versus being filled off the vehicle
- ☐ **C)** All of the above
- ☐ **D)** None of the above

21. What is a safe haven?

- ☐ **A)** A safe place to dump hazardous materials
- ☐ **B)** A place that has been approved for parking unattended vehicles carrying explosives
- ☐ **C)** A slang term for the last stop at the end of one's transporting of hazardous materials
- ☐ **D)** All of the above

22. In the hazardous materials table, which column provides the hazardous material identification number for each material?

- ☐ **A)** Column 4
- ☐ **B)** Column 3
- ☐ **C)** Column 2
- ☐ **D)** Column 6

23. What is a placard used for?

- ☐ **A)** To make other drivers stay away up to a minimum of 20 feet
- ☐ **B)** To warn those with children to drive in the other lane
- ☐ **C)** To communicate risk
- ☐ **D)** For fun purposes

24. What is a cargo tank?

- ☐ **A)** Bulk packaging attached to a vehicle
- ☐ **B)** Bulk packaging temporarily attached to a vehicle
- ☐ **C)** They are filled while they are off your vehicle and attached for transportation
- ☐ **D)** None of the above

25. What is the purpose of the shipper's certification on packaged materials?

- ☐ **A)** To make it as light as possible
- ☐ **B)** To make it easy for identification
- ☐ **C)** To make opening and closing of the package very easy
- ☐ **D)** For safety purposes

26. One of the following three hazard classes should not be placed in a temperature-controlled trailer

- ☐ **A)** Class 2.3 and 6
- ☐ **B)** Class. 1, 2.1 and 3
- ☐ **C)** Class 1, 3 and 6
- ☐ **D)** None of the above

27. Where is the best place to keep shipping documents that describe the hazardous material?

- ☐ **A)** Hidden in a safe place inside the tractor
- ☐ **B)** In a holder inside the driver's side door or sitting on the driver's seat
- ☐ **C)** In a locked glove compartment anytime you are out of the vehicle
- ☐ **D)** All of the above

28. One of the following hazardous classes uses a transport index to determine how much of it can be loaded on a single vehicle for transportation

- ☐ **A)** Class 4 (flammable solids)
- ☐ **B)** Class 3 (flammable liquids)
- ☐ **C)** Class 7 (radioactive materials)
- ☐ **D)** Class 2 (inflammable solids)

29. How many feet is one allowed to park from a bridge, tunnel, or building when carrying division 1.2 or. 1.3 materials?

- ☐ **A)** 300 feet
- ☐ **B)** 500 feet
- ☐ **C)** 600 feet
- ☐ **D)** 800 feet

30. What can be done in a scenario where you discover hazardous material leakage at a rest stop but there's no phone available?

- ☐ **A)** Send someone for help with all the necessary information
- ☐ **B)** Drive for help as quickly as possible
- ☐ **C)** Drive slowly and cautiously until you get to a phone
- ☐ **D)** None of the above

31. What does it mean if a "W" appears in column one of the hazardous material table?

- ☐ **A)** It identifies a proper shipping name that is used in international transportation
- ☐ **B)** It only applies when the material is being transported by water
- ☐ **C)** It only applies when the material will be exposed to air
- ☐ **D)** None of the above

32. How is the hazardous material table ordered?

- ☐ **A)** In numerical order by hazard class
- ☐ **B)** In alphabetical order by the proper shipping name
- ☐ **C)** In numerical order by identification number
- ☐ **D)** In Roman numerals

33. You are not allowed to use a hazard class name or ID number to describe...

- ☐ **A)** A non-hazardous material
- ☐ **B)** Hazardous waste
- ☐ **C)** The reportable quantity of hazardous material
- ☐ **D)** None of the above

34. What kind of materials are classified under hazard class 7?

- ☐ **A)** Radioactive materials
- ☐ **B)** Corrosive materials
- ☐ **C)** Flammable and combustible liquid
- ☐ **D)** Inflammable materials

35. The letter "G" in the hazardous material column means...

- ☐ **A)** Ground shipping
- ☐ **B)** Generic shipping name
- ☐ **C)** Geographical shipping name
- ☐ **D)** None of the above

36. A hazardous material's identification is determined by...

- ☐ **A)** The receiver
- ☐ **B)** The shipper
- ☐ **C)** The driver
- ☐ **D)** All of the above

37. If a shipment contains both hazardous materials, the shipping paper must...

- ☐ **A)** List the hazardous materials first
- ☐ **B)** Highlight the hazardous materials in contrasting colors
- ☐ **C)** All of the above
- ☐ **D)** None of the above

38. Who is responsible for finding out special routes needed to haul hazardous materials?

- ☐ **A)** The carrier
- ☐ **B)** The driver
- ☐ **C)** The shipper
- ☐ **D)** All of the above

39. The hazard class 6 includes...

- ☐ **A)** Poisons
- ☐ **B)** Infectious substances
- ☐ **C)** All of the above
- ☐ **D)** None of the the above

40. Examples of substances under hazard class 1 are...

- ☐ **A)** Gasoline, methanol
- ☐ **B)** Dynamite, fireworks
- ☐ **C)** Food flavorings
- ☐ **D)** None of the above

41. One of the following does not belong in hazard class 1 through 9

- ☐ **A)** Medicine, food flavoring
- ☐ **B)** Dynamite, fireworks
- ☐ **C)** Gasoline, methanol
- ☐ **D)** All of the above

42. Most identification numbers are preceded by the letters “UN” or “NA”. The proper shipping names that are associated with the letters “NA” are used where?

- ☐ **A)** Within the United States, and to and fro Canada
- ☐ **B)** Internationally
- ☐ **C)** All of the above
- ☐ **D)** None of the above

43. Containment rules are rules that _______

- ☐ **A)** Instruct the driver on how to unload hazardous materials
- ☐ **B)** Instruct the driver on how to load hazardous materials
- ☐ **C)** Instruct the driver on how to transport hazardous materials
- ☐ **D)** All of the listed answers

44. There is a hazard warning label placed by shippers on all packages of hazardous materials. What shape is the hazard warning label on the majority of packages?

- ☐ **A)** Oval
- ☐ **B)** Circle
- ☐ **C)** Diamond
- ☐ **D)** Rectangle

45. The letters of the shipping name must be, at minimum, how tall on portable tanks with capacities of more than 1,000 gallons?

- ☐ **A)** One inch.
- ☐ **B)** Three inches
- ☐ **C)** Five inches
- ☐ **D)** Two inches

46. Upon discovering a fire while you are transporting hazardous materials and the trailer doors feel hot, you should ______

- ☐ **A)** Try to fight the fire regardless of whether the cargo is already on fire
- ☐ **B)** Not open the doors
- ☐ **C)** Fan the fire
- ☐ **D)** Open the doors

47. Bulk packaging is...

- ☐ **A)** A single container with a capacity of 50 gallons
- ☐ **B)** Triple containers with each not exceeding a capacity of 25 gallons
- ☐ **C)** A single container with a capacity of 119 gallons or more of hazardous materials
- ☐ **D)** Double containers with a capacity of 100 gallons

48. When transporting hazardous waste, you need to ______

- ☐ **A)** Carry a Uniform Hazardous Waste Manifest in the truck
- ☐ **B)** Have the word “WASTE” before the name of the material on the bills
- ☐ **C)** Sign a uniform Hazardous Waste Manifest
- ☐ **D)** All of the answers above

49. What is the primary purpose of placards on a vehicle transporting hazardous materials?

- ☐ **A)** To make the vehicle more visible.
- ☐ **B)** To indicate the type of hazard the materials present.
- ☐ **C)** For advertising purposes.
- ☐ **D)** To show the vehicle's destination.

50. What must a driver check for before transporting hazardous materials?

- ☐ **A)** The weather forecast.
- ☐ **B)** The vehicle's fuel level.
- ☐ **C)** Proper labeling and securement of the materials.
- ☐ **D)** Traffic conditions.

51. In case of a hazardous material spill, what is the first action the driver should take?

- ☐ **A)** Continue driving to the nearest service station.
- ☐ **B)** Notify the appropriate authorities and follow emergency procedures.
- ☐ **C)** Clean the spill immediately.
- ☐ **D)** Repackage the materials.

6_ Double/Triple Trailers Practice Test

This full-length CDL Double/Triple Trailers practice test contains 46 questions. This test contains more CDL Double/Triple Trailers practice questions than you'll probably need to pass the Double/Triple Trailers endorsement, but why take any chances? Answering 46 practice questions will prepare you better than answering a mere 10 questions!

1. Whenever you're pulling more than one trailer, which should be behind the tractor?

- ☐ **A)** The heaviest trailer
- ☐ **B)** The shortest trailer
- ☐ **C)** The lightest trailer
- ☐ **D)** None of the above

2. One of the following is true about quick steering movement and double/triple vehicles

- ☐ **A)** They flip over from quick steering movement more easily than other vehicles
- ☐ **B)** Counter steering is easier in double/triple vehicles than in other vehicles
- ☐ **C)** Applying the brakes at the same time will let you perform the quick steering movement
- ☐ **D)** None of the above

3. How can one make sure to apply air to the second trailer?

- ☐ **A)** Watch each trailer air gauge for a drop of 30 psi
- ☐ **B)** Go to the rear of the second trailer and open the emergency line shut off valve
- ☐ **C)** Apply the hand valve at 10mph and stop at the same distance as a truck with the trailer at 5mph
- ☐ **D)** None of the above

4. Some trucks have convex or spot mirrors. What are they used for?

☐ **A)** They make things look smaller and farther than they really are

☐ **B)** They need not be checked as often as flat mirrors

☐ **C)** They make things look larger and closer than they really are

☐ **D)** All of the above

5. Driving a trailer or truck with a double or triple trailer requires the driver to...

☐ **A)** Allow for more following distance than with smaller vehicles

☐ **B)** Take care in special conditions

☐ **C)** All of the above

☐ **D)** None of the above

6. One of the following statements about steering is true

☐ **A)** You should apply brakes while turning

☐ **B)** It is not necessary to put both hands on the steering wheel

☐ **C)** The sharper you turn, the more likely the vehicle will skid or rollover

☐ **D)** None of the above is true

7. What will happen if the pintle hook is unlocked while the dolly is still under the second trailer?

☐ **A)** The dolly tow bar may fly up

☐ **B)** The air lines will rupture

☐ **C)** Nothing will happen

☐ **D)** None of the above

8. To check if the converter dolly is coupled to the rear trailer, one should...

☐ **A)** Pull against the pin of the second trailer

☐ **B)** Release spring brake and check the movements

☐ **C)** Pull forward and do a visual check of the coupling

☐ **D)** None of the above

9. One of the following statements about driving through a curve is false

☐ **A)** Slow down before entering the curve

☐ **B)** Brake and downshift in the curve

☐ **C)** Speed up as soon as you are out of the curve

☐ **D)** None of the above

10. One of the following is true about handling doubles and triples

☐ **A)** Triple bottom rigs can stop quicker than a 5-axle tractor semi-trailer due to off tracking

☐ **B)** Sudden movements with the steering wheel can result in a tipped-over rear trailer

☐ **C)** All of them is true

☐ **D)** None of them is true

11. When you want to hook your combination vehicle to a second trailer that has no spring brakes and wheel chocks, you should...

☐ **A)** Supply air to the trailer air system with the tractor and disconnect the emergency line

☐ **B)** Make sure the trailer will roll freely when coupling

☐ **C)** Hook the trailer electric cord to a portable generator for braking power

☐ **D)** All of the above

12. One of these statements is true

☐ **A)** Empty rigs are more likely to turn over than fully loaded rigs

☐ **B)** When driving a straight truck, you should look further ahead than when you are pulling doubles

☐ **C)** You should drive a double or triple very smoothly to avoid a rollover

☐ **D)** All are true

13. When driving a 100-feet twin trailer combination at 50mph and the road is dry with good visibility, you should leave at least ___ seconds of space ahead of you

- ☐ **A)** 11
- ☐ **B)** 9
- ☐ **C)** 10
- ☐ **D)** 7

14. What is a converter dolly?

- ☐ **A)** It is an electronic device that connects the electric power from a semi-trailer to the rear of the tractor-trailer
- ☐ **B)** It is a coupling device that connects a semi-trailer to the rear of a tractor-trailer
- ☐ **C)** All of the above
- ☐ **D)** None of the above

15. When pulling a trailer at under 40mph, how much following distance should you maintain behind the vehicle in front of you?

- ☐ **A)** 3 seconds
- ☐ **B)** 1 second
- ☐ **C)** 2 seconds
- ☐ **D)** 4 seconds

16. With the hand valve on, a driver should test the trailer brakes by opening the service line valve at the rear of the rig. When this is done, what will be heard?

- ☐ **A)** The emergency line valve open
- ☐ **B)** Air escape from the open valve
- ☐ **C)** Service brake slowly move into the fully applied position
- ☐ **D)** None of the above

17. When doing the walk-around inspection of a vehicle, you should check landing gear to be sure that....

☐ **A)** It is fully raised

☐ **B)** It isn't damaged

☐ **C)** All of the above

☐ **D)** None of the above

18. Drivers should make sure that the converter dolly air tank drains are ____ and the pintle hook is _____

☐ **A)** Closed: latched

☐ **B)** Open: free

☐ **C)** Open: latched

☐ **D)** Closed: open

19. In a vehicle that loads twin semi-trailers, one should lower its landing gear until _____ while uncoupling the rear semi-trailer

☐ **A)** It touches the ground

☐ **B)** It is enough to take the weight off the dolly

☐ **C)** It lifts the trailer off the dolly

☐ **D)** None of the above

20. For coupled doubles and triples, what is the correct position for the converter dolly air tanks petcocks?

☐ **A)** Closed

☐ **B)** Open

☐ **C)** Open, except for the dolly connected to the last trailer

☐ **D)** Free

21. When driving double trailers and you have to use your brakes to avoid a crash, you should _____ for emergency brakes:

☐ **A)** Use controlled or stab braking

☐ **B)** Use trailer brakes only

☐ **C)** Push the brake pedal as hard and hold it there

☐ **D)** Steering wheel

22. When pulling doubles and a set of trailer wheels go into a skid, which of the following is likely to happen?

☐ **A)** The rig will continue to move in a straight line no matter how the steering wheel is turned

☐ **B)** The rig will stay in a straight line

☐ **C)** The rig may jackknife

☐ **D)** None of the above

23. ______ are gas or hydraulic devices that cushion the vehicle ride and stabilize the vehicle

☐ **A)** Springs

☐ **B)** Brakes

☐ **C)** Shock absorbers

☐ **D)** Water pump

24. Part of the vehicle inspection test especially if you have a cargo lift is to...

☐ **A)** Check for leaks and damages

☐ **B)** Explain the process of checking it for correct operation

☐ **C)** All of the above

☐ **D)** None of the above

25. Which of the gauges indicates that the alternator or generator is charging?

- ☐ **A)** Temperature
- ☐ **B)** Voltmeter
- ☐ **C)** Oil pressure
- ☐ **D)** Spring brakes

26. In order to improve vision during stormy weather, what must be secure, undamaged, and operating smoothly?

- ☐ **A)** The lightning indicators
- ☐ **B)** Windshield wipers and washers
- ☐ **C)** The emergency equipment
- ☐ **D)** None of the above

27. In the case that the coolant overflow is not part of the pressurized system, check the coolant level by...

- ☐ **A)** Removing the container cap and checking the visible
- ☐ **B)** Inspecting the reservoir
- ☐ **C)** Doing either A or B
- ☐ **D)** None of the above

28. What is a shipper's certification?

- ☐ **A)** A shipper's statement that the product is not safe to inhale
- ☐ **B)** The shipper's certification that the shipment has been prepared according to applicable rules
- ☐ **C)** The shipper's description of the hazardous materials, including the proper shipping name, the class of the hazardous materials, the ID number, and the packing group
- ☐ **D)** All of the above

29. How much space should be between the upper and fifth lower wheel especially during the inspection of the coupling of a converter dolly to the rear trailer?

- ☐ **A)** None
- ☐ **B)** ½ and ¾ inch space
- ☐ **C)** It depends on the load
- ☐ **D)** All of the above

30. When transporting class 1.1, 1.2, or 1.3 materials, you should use a placard if they exceed...

- ☐ **A)** 440 pounds
- ☐ **B)** The reportable quantity amount
- ☐ **C)** Any amount
- ☐ **D)** There is no need to use a placard for such materials

31. In the case that someone is attending to a placarded vehicle that is parked, that person should...

- ☐ **A)** Be able to move the vehicle if necessary
- ☐ **B)** Either be in the vehicle, awake, or within 100 feet of the vehicle
- ☐ **C)** Both A and B
- ☐ **D)** None of the above

32. One must stop at a railroad crossing if their vehicle:

- ☐ **A)** Must display specific placards as required
- ☐ **B)** Transporting any amount of chlorine
- ☐ **C)** All of the above
- ☐ **D)** None of the above

33. Before connecting a converter dolly to the second or third trailer, you should check the height of the trailer. The trailer height is right if...

- ☐ **A)** The kingpin is resting on the fifth wheel
- ☐ **B)** The trailer will be raised slightly when the converter dolly is backed under it
- ☐ **C)** The center of the king pin lines up with the locking jaws
- ☐ **D)** None of the above

34. When transporting materials from hazard table one, which rule applies?

- ☐ **A)** A placard is required for any amount
- ☐ **B)** A placard is required if the total amount transported is up to 1,001 pounds or more
- ☐ **C)** You may use dangerous placards if you have 1,0001 pounds or more of two different classes
- ☐ **D)** None of the above

35. The official CDL manual section on double and triple trailers usually describes how to couple and uncouple...

- ☐ **A)** Every type of combination
- ☐ **B)** Only the more common types of combination
- ☐ **C)** Only doubles
- ☐ **D)** None of the above

36. Parking brakes should be used whenever you park excluding what situation?

- ☐ **A)** The brakes are very hot
- ☐ **B)** All of the listed answers
- ☐ **C)** You have just come down a steep grade
- ☐ **D)** The brakes are wet in freezing temperatures

37. A typical low air pressure warning signal is in what color?

- ☐ **A)** Purple
- ☐ **B)** Blue
- ☐ **C)** Yellow
- ☐ **D)** Red

38. Vehicles with ABS have malfunction lamps that are what color?

- ☐ **A)** Yellow
- ☐ **B)** White
- ☐ **C)** Green
- ☐ **D)** Blue

39. What should be done first to test the trailer emergency brakes?

- ☐ **A)** Push in the trailer air supply control
- ☐ **B)** Pull harshly on the trailer
- ☐ **C)** Pull gently on the trailer with the tractor
- ☐ **D)** Charge the trailer air brake system

40. Why should the use of automatic transmission when going into a lower gear at high speed be avoided?

- ☐ **A)** None of the listed answers
- ☐ **B)** It will always completely destroy the transmission
- ☐ **C)** It could damage the transmission and lead to a loss of the engine's braking effect
- ☐ **D)** This is a trick question. You should force an automatic transmission into a lower gear while you are driving at a high speed as doing so will help you maintain control of the vehicle.

41. What differentiates a wedge brake from other types of brakes?

☐ **A)** None of the listed answers

☐ **B)** It stops the vehicle completely

☐ **C)** Activated by Air Pressure

☐ **D)** The brake chamber push rod pushes a wedge directly between the ends of two brake shoes

42. Why do doubles and triples require more space in comparison to other types of commercial vehicles?

☐ **A)** They have more height and length than tractors

☐ **B)** They have more length and swing from side to side as a result

☐ **C)** They have more length and cannot be stopped or turned abruptly

☐ **D)** It is a requirement by the law

43. Why do doubles and triples have a higher probability of skidding in slippery conditions, bad weather, and during mountain driving?

☐ **A)** There may not be any brakes on the front wheels

☐ **B)** There is greater length and more dead axles to be pulled

☐ **C)** The brakes are more prone to failure

☐ **D)** The drive wheels only secure the pulling unit

44. What is crucial to remember when making a turn with a double or triple trailer?

☐ **A)** Always use the handbrake while turning

☐ **B)** Ensure the rear trailer follows the path of the front trailer

☐ **C)** Make wider turns than usual

☐ **D)** Accelerate rapidly while turning

45. What is the primary purpose of a converter gear on a double or triple trailer?

- ☐ **A)** To convert air pressure for braking
- ☐ **B)** To connect multiple trailers
- ☐ **C)** To assist in reversing maneuvers
- ☐ **D)** To provide additional cargo space

46. What should be your main concern when pulling triples on a wet road?

- ☐ **A)** The increased stopping distance
- ☐ **B)** The potential for the rear trailer to skid
- ☐ **C)** Difficulty in steering
- ☐ **D)** Reduced visibility

7_Combination Vehicles Practice Test

This full-length CDL Combination Vehicles practice test contains 47 questions. This test contains more CDL Combination Vehicles practice questions than you'll probably need to pass the Combination Vehicles endorsement, but why take any chances? Answering 47 practice questions will prepare you better than answering a mere 10 questions!

1. What is the best way to tell if your trailer has started to skid?

- ☐ **A)** Seeing it in the mirror
- ☐ **B)** Feeling for pulling in the cab
- ☐ **C)** Hearing the wheels skidding

2. When inspecting the landing gear after uncoupling the trailer, where should the tractor be?

- ☐ **A)** Completely clear of the trailer
- ☐ **B)** With the tractor frame under the trailer
- ☐ **C)** The fifth wheel directly under the kingpin

3. There are two things the driver can do to prevent a roll-over- keeping the cargo as close to the ground as possible and...

- ☐ **A)** Going slowly around turns
- ☐ **B)** Making sure the vehicle brakes are properly adjusted
- ☐ **C)** Keeping the fifth wheel free play as tight as possible

4. After locking the king pin into the fifth wheel, how can you check the connection?

- ☐ **A)** Pull the tractor ahead gently with the trailer brake locked
- ☐ **B)** Pull forward 50 feet, then turn right and left
- ☐ **C)** Pull the tractor ahead sharply to release the trailer brake

5. If you have a semi-trailer, where should you put the front trailer tha supports it before driving away?

☐ **A)** Raised halfway with the crack handle secured in its bracket

☐ **B)** Fully raised with the crack handle secured in its bracket

☐ **C)** None of the above

6. The safest way to turn right from a two-way road is...

☐ **A)** Turning wide as you complete the turn

☐ **B)** Turning wide before starting the turn

☐ **C)** Backing up if there are vehicles in that lane if you have to cross to the opposing lane

7. While uncoupling, you should disconnect the electrical cable and...

☐ **A)** Coil it to keep it out of the way

☐ **B)** Hang it with the plug down

☐ **C)** Hang it with the plug up

8. More than half of truck drivers' deaths is as a result of...

☐ **A)** Speed

☐ **B)** Following too closely

☐ **C)** Roll-overs

9. Low-slung vehicles can be risky at railroad crossings because...

☐ **A)** They may take longer to stop

☐ **B)** They are more likely to get stuck on a railroad crossing

☐ **C)** They are more likely to jackknife on the uneven ground

10. Which of the following is true regarding the use of drugs when it comes to driving?

☐ **A)** Prescription or non-prescription drugs are not allowed for any reason

☐ **B)** Prescription drugs are allowed as long as the doctor says it will not affect your driving ability

☐ **C)** Speed is allowed as long as you are taking drugs to stay awake

11. One of the following statements is true about tires in hot weather

☐ **A)** You should inspect your tires every two hours or every 100 miles when driving in very hot weather

☐ **B)** You should check tire mounting and air pressure before driving

☐ **C)** You should bleed a small amount of air to keep tire pressure steady

12. Some suspension system defects are...

☐ **A)** Leaking shock absorbers

☐ **B)** Broken leaves in a leaf spring and a cracked spring hanger

☐ **C)** All of the above

13. If the trailer has an Antilock Braking System but the tractor doesn't, what is likely to happen?

☐ **A)** The trailer is likely to swing out

☐ **B)** The trailer is still less likely to swing out

☐ **C)** None of the above

14. When you are pulling doubles or triples, the shut-off valves should always be...

☐ **A)** Open, except for those on the last trailer

☐ **B)** Open

☐ **C)** Open, except for those on the first trailer

15. After connecting the air lines but before backing under the trailer, one should...

- ☐ **A)** Make sure that the trailer brakes are off
- ☐ **B)** Supply air to the trailer system and then pull out the air supply knob
- ☐ **C)** Walk around the rig to make sure that it is clear

16. What is a crack-the-whip effect according to one of these statements?

- ☐ **A)** If you make a sudden movement of your steering wheel, your trailer will tend to swing out. The rearmost trailer will swing out the most and will be more likely to roll over
- ☐ **B)** If a sudden movement is made from your steering wheel, the tractor will tend to rock and sway
- ☐ **C)** When your trailer is half full and loaded in the front of the trailer, sudden movement of your steering wheel will cause your cargo to slide

17. If your vehicle gets stuck on a railroad track, you should...

- ☐ **A)** Get away from your vehicle and call 911
- ☐ **B)** Radio in for assistance
- ☐ **C)** Honk the horn loudly and call 911

18. Before uncoupling, you should...

- ☐ **A)** Call in for any special requests
- ☐ **B)** Make sure your vehicle is parked on a level surface, the ground is solid and can support the weight of the trailer
- ☐ **C)** Make sure the trailer has enough air supply for the brakes to hold

19. Combination vehicles take longer to stop when they are empty because...

- ☐ **A)** There is less traction
- ☐ **B)** You are likely to brake as hard
- ☐ **C)** The center of gravity is lower

20. If you must drive on slippery roads, what is a good thing to do in such situations?

- ☐ **A)** Slow down gradually
- ☐ **B)** Use smaller following distance
- ☐ **C)** Apply brakes during turns
- ☐ **D)** None of the above

21. One can confirm if air is going to all brakes in the trailer by...

- ☐ **A)** Opening the emergency line shut off-valves at the rear of the last trailer and listening for air escaping
- ☐ **B)** Opening the emergency line shut-off valves and the service line valve at the rear of the last trailer and listening for air escaping each time
- ☐ **C)** Confirming that the air pressure is at a normal level

22. To prevent a roll-over, cargo should be...

- ☐ **A)** Spread evenly across the rig as low as possible
- ☐ **B)** Stacked closest to the door
- ☐ **C)** Stacked as close to the front of the rig as possible.

23. When testing ONLY the trailer service brakes, which of the options below about the air valves is correct?

- ☐ **A)** Parking Brake: push in; Trailer air supply: in / open;
- ☐ **B)** Parking Brake: push in; Trailer air supply: out / closed;
- ☐ **C)** Parking Brake: pull out; Trailer air supply: out / closed;

24. How should you test the tractor-trailer connection for security?

- ☐ **A)** Pull gently forward in low gear against the locked trailer brakes, then inspect the coupling
- ☐ **B)** Rock the trailer back and forth with the trailer brakes locked
- ☐ **C)** Put the tractor in gear and pull ahead with a sharp jerk

25. You should not use the trailer hand valve while driving because...

- ☐ **A)** You should only use the parking brake
- ☐ **B)** Of the danger of making the trailer skid
- ☐ **C)** It won't work as well as the foot brake

26. The air leakage rate for a combination vehicle (engine off, brakes on) should not be more than ____ psi per minute

- ☐ **A)** 2
- ☐ **B)** 3
- ☐ **C)** 4

27. Why should you lock the tractor gladhands to each other or to the storage bracket when you are not towing a trailer?

- ☐ **A)** Because the connected brake circuit becomes a backup air tank
- ☐ **B)** Because it will keep dirt and water out of the lines
- ☐ **C)** Because if you do not, you can never build system pressure

28. The hand valve should be used...

- ☐ **A)** Only with the foot brake
- ☐ **B)** To test the trailer brakes
- ☐ **C)** Only when the trailer is fully loaded

29. If the brakes do not release when you push the trailer air supply valve, you should...

- ☐ **A)** Check the electrical cables
- ☐ **B)** Cross the air lines
- ☐ **C)** Check the air line connections

30. In general, the higher your truck's center of gravity, the...

- ☐ **A)** Easier it is to turn over
- ☐ **B)** Easier it is to turn around corners
- ☐ **C)** More stable it is when turning

31. When you test the tractor protection valve, the red air supply control knob should go into the ________ position

- ☐ **A)** Emergency
- ☐ **B)** Normal
- ☐ **C)** Neutral

32. When the front trailer supports are raised, and the trailer is resting on the tractor, you should make sure that...

- ☐ **A)** There is enough clearance between the tops of the tractor tires and the nose of the trailer
- ☐ **B)** There is enough clearance between the tractor frame and the landing gear
- ☐ **C)** Both answers are correct

33. To test the tractor protection valve, charge the trailer air brake system, turn off the engine, and...

- ☐ **A)** Keep pressing the brake pedal firmly
- ☐ **B)** Flash your high-beam headlights on and off several times
- ☐ **C)** Step on and off the brake pedal several times

34. One of these vehicles off-tracks the most

- ☐ **A)** A 5-axle tractor towing a 45-foot trailer
- ☐ **B)** A 5-axle tractor towing a 42-foot trailer
- ☐ **C)** A 5-axle tractor towing a 52-foot trailer

35. Describe what the trailer air supply control does

- ☐ **A)** It is a device used to keep the trailer behind the tractor
- ☐ **B)** It is a six-sided yellow knob used to control the tractor protection valve
- ☐ **C)** It is used to supply the trailer with air, shut the air off, and apply the trailer emergency brakes

36. Why should you connect the glad hands to the dummy couplers if your vehicle is equipped with them?

☐ **A)** It will keep dirt and water out of the lines

☐ **B)** If you don't, you will never build system pressure

☐ **C)** The connected brake circuit becomes a backup air tank

37. In what position should the tractor protection valve control be placed in order to test the trailer air brakes when coupling a tractor-semitrailer?

☐ **A)** Up

☐ **B)** Down

☐ **C)** Normal

38. Trucks roll over more easily when fully loaded and are...

☐ **A)** Five times more likely to roll over during a crash than empty rigs

☐ **B)** Two times more likely to roll over during a crash than empty rigs

☐ **C)** Ten times more likely to roll over during a crash than empty rigs

39. If the driver crosses the air lines when hooking up to an old trailer, what will happen?

☐ **A)** You can drive away if the trailer has no spring brakes, but you will not have trailer brakes

☐ **B)** Instead of the trailer brakes, the hand valve will apply the tractor brakes

☐ **C)** Instead of the air brakes, the brake pedal will work the trailer spring brakes

40. Why should you be sure that the fifth wheel plate is greased as required?

☐ **A)** To ensure good electrical connections

☐ **B)** To prevent steering problems

☐ **C)** To reduce heat and noise

41. What is a tractor jackknife?

☐ **A)** When the drive tires on the tractor get locked up and the tractor spins out sideways as the trailer continues pushing forward

☐ **B)** When the drive tires on the trailer get locked up and the trailer spins out sideways as the tractor continues pushing forward

☐ **C)** Neither of the above

42. After you have uncoupled the trailer and are inspecting the trailer supports, in what gear should the tractor engine be?

☐ **A)** Neutral

☐ **B)** High reverse

☐ **C)** Low reverse

43. You should line up ____ as you get ready to back under the semi-trailer

☐ **A)** About 15 degrees away from the line of the trailer

☐ **B)** So that the king pin engages the driver's side locking jaw first

☐ **C)** Directly in front of the trailer

44. During uncoupling, you should disengage the electrical cable and...

☐ **A)** Coil it to keep it out of the way

☐ **B)** Hang it with the plug down

☐ **C)** Hang it with the plug

45. When coupling a tractor to a semi-trailer, you should:

☐ **A)** Connect the air lines before backing under the trailer.

☐ **B)** Back under the trailer before connecting the air lines.

☐ **C)** Make sure the trailer brakes are locked.

☐ **D)** Ensure the kingpin is damaged.

46. How should you check that the trailer is securely coupled to the tractor?

- ☐ **A)** Pull gently forward in low gear while the trailer brakes are still locked.
- ☐ **B)** Drive in a circle and check in the mirrors.
- ☐ **C)** Accelerate quickly and check in the mirrors.
- ☐ **D)** Listen for noise from the coupling.

47. What should you do if the low air pressure warning comes on in a combination vehicle?

- ☐ **A)** Continue driving and check the brakes when convenient.
- ☐ **B)** Stop and safely park as soon as possible.
- ☐ **C)** Increase your speed to build up air pressure.
- ☐ **D)** Pump the brakes to increase pressure.

8_School Bus Practice Test

This full-length CDL School Bus practice test contains 52 questions. This test contains more CDL School Bus practice questions than you'll probably need to pass the School Bus endorsement, but why take any chances? Answering 52 practice questions will prepare you better than answering a mere 10 questions!

1. In a school bus, what do drivers use the overhead inside rearview mirror for?

- ☐ **A)** To monitor the passengers actively inside the bus
- ☐ **B)** To monitor traffic that approaches
- ☐ **C)** To see the blind spot immediately behind the bus
- ☐ **D)** All of the above

2. When loading students and you cannot account for a student, what should be done?

- ☐ **A)** Look in your mirror for the student
- ☐ **B)** Secure the bus, take the key and check around for the student
- ☐ **C)** Ask other students if they saw that particular student
- ☐ **D)** None of the above

3. The blind spot behind the school bus may extend up to...

- ☐ **A)** 200 feet
- ☐ **B)** 400 feet
- ☐ **C)** 100 feet
- ☐ **D)** 500 feet

4. What is a passive railroad crossing?

- ☐ **A)** A railroad crossing that doesn't have a crossbuck sign
- ☐ **B)** A railroad crossing that doesn't have any type of traffic control design
- ☐ **C)** A railroad crossing at which one is not required to stop
- ☐ **D)** All of the above

5. During a pre-trip vehicle inspection, drivers should make sure that...

- ☐ **A)** Emergency exit signs are working
- ☐ **B)** Emergency exit signs are undamaged and ABS is closely secure from the inside
- ☐ **C)** Both of the above are correct
- ☐ **D)** None of the above

6. If your bus has a manual transmission, you should never ________

- ☐ **A)** Accelerate
- ☐ **B)** Honk your horn
- ☐ **C)** Change gears while crossing railroad tracks
- ☐ **D)** Slow down your vehicle before approaching a railroad crossing

7. As a bus driver, you need to avoid...

- ☐ **A)** Inspecting your bus before you drive
- ☐ **B)** Fueling the bus in a closed building while there are passengers on board
- ☐ **C)** Maintaining the speed limit
- ☐ **D)** All of the listed answers

8. If an object is dropped by a student in front of the bus, what should you tell the students to do as the bus driver?

- ☐ **A)** To attempt to grab your attention to retrieve the object
- ☐ **B)** All of the listed answers
- ☐ **C)** To leave any dropped object behind
- ☐ **D)** To move to a point of safety out of the danger zones

9. What should you tell your students to prevent dangerous situations during their loading and unloading?

☐ **A)** Tell students to leave behind any dropped object, step out of the danger zones, and call your attention so that the object can be retrieved

☐ **B)** Tell students that any dropped objects can be picked as long as they are not under the bus

☐ **C)** Tell students not to use the handrail when exiting the bus because they could be caught doing so

☐ **D)** None of the above

10. What is the best way to control student safety during an emergency (e.g. the bus engine failing)?

☐ **A)** Having the students evacuate the bus

☐ **B)** gnoring the students

☐ **C)** Accelerating and driving as quickly as possible to a safe destination

☐ **D)** Keeping the students on the bus if possible

11. The recommended safe distance to stop before a drawbridge is _______

☐ **A)** 100 feet

☐ **B)** 50 feet

☐ **C)** 150 feet

☐ **D)** 200 feet

12. When refueling your bus, it is extremely important that you remember to never __________

☐ **A)** Pay for gas

☐ **B)** Refuel your bus in a closed building with passengers on board

☐ **C)** Be polite

☐ **D)** None of the listed answers

13. Which of the following parts of your bus needs to be in safe working condition before driving?

☐ **A)** Each handhold and railing

☐ **B)** Emergency exit handles

☐ **C)** Signaling devices

☐ **D)** All of the above

14. In the case of a fire outbreak on the bus, where should you lead students?

☐ **A)** Upwind of the bus

☐ **B)** To the left of the bus

☐ **C)** To the right of the bus

☐ **D)** Downwind of the bus

15. Remember, as a school bus driver, you should __________ before entering traffic flow

☐ **A)** All of the listed answers

☐ **B)** Allow congested traffic to disperse

☐ **C)** Call the parents and tell them when their children will arrive

☐ **D)** Open the bus doors

16. Students may drop an object near the bus during loading and unloading, you should _______

☐ **A)** All of the listed answers

☐ **B)** Watch carefully as they enter and exit the bus

☐ **C)** Ignore them

☐ **D)** Leave the students behind

17. The bus' alternating flashing amber warning lights must be activated in accordance with state laws or at least _____ before the school bus stop

- ☐ **A)** 100 feet
- ☐ **B)** 25 feet
- ☐ **C)** 200 feet
- ☐ **D)** 50 feet

18. The driver must evacuate the bus when _________

- ☐ **A)** The bus is on fire
- ☐ **B)** All of the listed answers
- ☐ **C)** The bus is stalled on a railroad-highway crossing
- ☐ **D)** There is an imminent danger of collision

19 If the bus is so equipped, how many seconds before reaching a stop do school bus drivers need to switch on their alternating flashing amber warning lights?

- ☐ **A)** 5-10 seconds
- ☐ **B)** 60 seconds
- ☐ **C)** 15-20 seconds
- ☐ **D)** 6-9 seconds

20. Who should you reach out to if you are having difficulty driving in high winds?

- ☐ **A)** No one
- ☐ **B)** The students' parents
- ☐ **C)** Your dispatcher
- ☐ **D)** Your loved ones

21. How many versions of the Class A Pre-trip Inspection test are there?

☐ **A)** Two

☐ **B)** Eight

☐ **C)** Three

☐ **D)** Four

22. After you have finished the route, you should _________

☐ **A)** None of the listed answers

☐ **B)** Conduct a post-trip inspection

☐ **C)** Call your loved ones

☐ **D)** Fill your gas tank

23. It's important for bus drivers to understand...

☐ **A)** What students do when crossing the street in front of the bus

☐ **B)** What students do when exiting the school bus

☐ **C)** Students might not always do what they are supposed to do

☐ **D)** All of the above

24. What should you do if there is no police officer and you believe the signal at an active railroad crossing is malfunctioning?

☐ **A)** Call your dispatcher

☐ **B)** Contact the school officials first

☐ **C)** Turn around and find another route

☐ **D)** Avoid the intersection

25. School bus drivers need to concentrate while they are driving and while the students are loading and unloading the bus. If a student exhibits a behavior problem that distracts you from this task, what should you do?

☐ **A)** Ignore the problem as much as you can and report the distracting child later on

☐ **B)** None of the listed answers

☐ **C)** Hold on until the students unloading get off the bus safely and move away or pull over to resolve the problem

☐ **D)** Discharge the student with the behavioral problem at a safe spot first. Then you should drop off the other students and then deal with the student after you have finished your route

26. Without _______ from the appropriate school district official, you should never change the location of a bus stop

☐ **A)** None of the listed answers

☐ **B)** A signed paycheck

☐ **C)** Written approval

☐ **D)** Verbal agreement

27. In preparing for your CDL Skills test, which of the following parts should you be able to identify?

☐ **A)** Emergency equipment

☐ **B)** Air brakes

☐ **C)** Exhaust system

☐ **D)** All of the listed answers

28. To perform a proper inspection of suspension mounts, you need to ______

☐ **A)** Inspect the mounts at each point where they are secured to the vehicle frame and axle(s)

☐ **B)** Look for cracked spring hangers

☐ **C)** Check for missing bolts

☐ **D)** All of the above

29. Which of the following is a reason why it is so important for the bus driver to inspect the bus' interior at the end of their shift?

☐ **A)** To collect and catalog all items left behind on their bus

☐ **B)** To find loose change that will help them pay for it

☐ **C)** To make sure they can report necessary repairs to a mechanic before they drive the bus again

☐ **D)** None of the above

30. What is the standee line in a bus?

☐ **A)** A two-inch line on the floor (or some other marking) that indicates where passengers cannot stand

☐ **B)** The number of standing spots you have on your bus

☐ **C)** The line of passengers standing at the bus stop

☐ **D)** Anywhere passengers are standing

31. During an emergency, remember that you can possibly evacuate through _______

☐ **A)** The window

☐ **B)** The rear door

☐ **C)** The side door

☐ **D)** All of the above

32. The _______ establishes the official routes and official school bus stops

☐ **A)** The school district

☐ **B)** The President of the United States

☐ **C)** Each school bus driver determines their routes and stops

☐ **D)** The parents of the children riding on the school bus

33. The Department of Transportation requires that ABS be on ________

- ☐ **A)** All of the listed answers
- ☐ **B)** Truck tractors with air brakes built on or after March 1, 1997
- ☐ **C)** Buses built on or after March 1, 1998
- ☐ **D)** Hydraulically-braked trucks and buses with a gross vehicle weight rating of 10,000 lbs. or more built on or after March 1, 1999

34. Which of the following would be an emergency situation that a school bus driver may encounter?

- ☐ **A)** A crash
- ☐ **B)** A student's medical emergency
- ☐ **C)** All of the listed answers
- ☐ **D)** An electrical fire in the engine compartment of a school bus

35. Before leaving the unloading area, be certain...

- ☐ **A)** The students are standing
- ☐ **B)** No students are returning to the bus
- ☐ **C)** The doors are open
- ☐ **D)** You have a tank full of gas

36. If you need to back up a bus without a lookout, you should...

- ☐ **A)** Use your mirrors and back up quickly
- ☐ **B)** Back up quickly and pray
- ☐ **C)** Set the parking brake, turn off the motor, take the keys with you and walk to the rear of the bus to determine whether the way is clear
- ☐ **D)** None of the above

37. If your fifth wheel plate is not well lubricated, this could cause...

- ☐ **A)** Nothing
- ☐ **B)** Steering problems
- ☐ **C)** None of the listed answers
- ☐ **D)** Less air to reach the emergency line

38. For school bus drivers, the danger zone is...

- ☐ **A)** A blind corner
- ☐ **B)** A forbidden street
- ☐ **C)** The front of a police station
- ☐ **D)** The area on all sides of a school bus

39. When trying to decide if the school bus should be evacuated during an emergency, which of these questions do you not need you ask yourself?

- ☐ **A)** Is there a chance the bus could be hit by other vehicles?
- ☐ **B)** Is there a fire?
- ☐ **C)** Is there a smell of leaking fuel?
- ☐ **D)** All of the above

40. When securing the bus during an evacuation, remember to...

- ☐ **A)** Place transmission in "Park"
- ☐ **B)** Set parking brakes
- ☐ **C)** Activate hazard-warning lights
- ☐ **D)** All of the above

41. When the red flashing lights on an eight-light system bus are activated, what does it signify to other drivers?

- ☐ **A)** That the school bus is stopped for a red light
- ☐ **B)** That the school bus is stopped for a train crossing
- ☐ **C)** That the school bus is stopped because it has broken down
- ☐ **D)** That the school bus is stopped to pick up or discharge pupils

42. If you're operating a bus on a divided highway or a highway that has up to or above four lanes, which of the following statements below describes the procedure for the proper pick-up and discharge of students?

☐ **A)** The students can only be picked up or dropped off on the resident side of the road

☐ **B)** The students can be dropped off or picked up on either side of the highway

☐ **C)** The students must be picked up and dropped off in the center of the road

☐ **D)** No answers are correct

43. The "golden rule" for school buses when they approach or cross railroad tracks is...

☐ **A)** Stop, go, and stop

☐ **B)** Stop, look, and listen

☐ **C)** Stop, look, and turn on the emergency flashers

☐ **D)** None of the above

44. When under the age of 21 years old, a driver will be considered intoxicated at...("Tolerance levels vary depending on the state one lives in, so it's important to check and refer to the regulations of one's state")

☐ **A)** Zero tolerance

☐ **B)** More than .02 but less than .10 BAC

☐ **C)** .10 BAC like any other driver

☐ **D)** None of the above

45. You should maintain your position _________ of the rightmost lane when you approach a bus stop on a multilane roadway

☐ **A)** Toward the right edge

☐ **B)** Toward the left edge

☐ **C)** In the center

☐ **D)** All of the above

46. Is the bus driver allowed to crack or open the service door prior to stopping to pick up a student?

- ☐ **A)** Yes
- ☐ **B)** No
- ☐ **C)** Maybe
- ☐ **D)** It depends on the situation

47. Which of the people listed below have the authority to remove or suspend a student from riding the bus?

- ☐ **A)** The school bus driver
- ☐ **B)** The school principal
- ☐ **C)** The school principal, superintendent, or the superintendent's designee
- ☐ **D)** The student's teacher

48. When are substances like alcohol, tobacco, and non-prescription drugs allowed to be used on the school bus?

- ☐ **A)** At any time
- ☐ **B)** Never
- ☐ **C)** Only when the bus driver says it is okay
- ☐ **D)** No answers are correct

49. The driver's seat should have a seat belt except when ______

- ☐ **A)** The driver has assumed responsibility for the risk
- ☐ **B)** It's very hot outside
- ☐ **C)** The driver's seat always needs to have a seat belt
- ☐ **D)** The bus was created before 1980

50. What is the most important action for a school bus driver to take at railroad crossings?

- ☐ **A)** Speed up to cross quickly.
- ☐ **B)** Stop, and look both ways.
- ☐ **C)** Honk to alert the train.
- ☐ **D)** Continue without stopping if no train is visible.

51. How far from the bus should students be when crossing the street in front of the bus?

- ☐ **A)** At least 10 feet.
- ☐ **B)** Right in front of the bus.
- ☐ **C)** 5 feet away.
- ☐ **D)** They should not cross in front of the bus.

52. What should a school bus driver do if a student is misbehaving on the bus?

- ☐ **A)** Ignore the behavior and continue driving.
- ☐ **B)** Stop the bus in a safe location and address the behavior.
- ☐ **C)** Let the student off the bus immediately.
- ☐ **D)** Call the police.

ANSWERS

1_General Knowledge Practice Test. Answers

1.____B
2.____B
3.____B
4.____D
5.____C
6.____D
7.____A
8.____B
9.____A
10.____A
11.____A
12.____D
13.____A
14.____D
15.____B
16.____D
17.____A
18.____B
19.____C
20.____D
21.____C
22.____A
23.____C
24.____B
25.____A
26.____A
27.____A
28.____D
29.____A
30.____B
31.____D
32.____A
33.____B
34.____D
35.____A
36.____C
37.____A
38.____A
39.____D
40.____D
41.____B
42.____B
43.____B

2_Pre-Trip Inspection Practice Test. Answers

1.____B
2.____C
3.____C
4.____A
5.____C
6.____B
7.____A
8.____A
9.____B
10.____A
11.____A
12.____A
13.____A
14.____A
15.____A
16.____C
17.____A
18.____B
19.____A
20.____B
21.____C
22.____B
23.____B
24.____B
25.____B
26.____A
27.____C
28.____A
29.____B
30.____C
31.____A
32.____B
33.____A
34.____A
35.____C
36.____B
37.____A
38.____D
39.____B
40.____A
41.____A
42.____A
43.____A
44.____D
45.____B
46.____B
47.____C

3_Air Brakes Practice Test. Answers

1._____D
2._____A
3._____B
4._____B
5._____C
6._____D
7._____B
8._____A
9_____C
10.____B
11.____D
12.____B
13.____C
14.____B
15.____A
16.____B
17.____D
18.____B
19.____B
20.____B
21.____A
22.____A
23.____D
24.____B
25.____A
26.____C
27.____D
28.____B
29.____D
30.____A
31.____C
32.____B
33.____A
34.____A
35.____B
36.____B
37.____C
38.____A
39.____C
40.____A
41.____B
42.____A
43.____A
44.____B
45.____B
46.____C
47.____A

4_Tanker Vehicles Practice Test. Answers

1._____A
2._____A
3._____C
4._____A
5._____C
6._____B
7._____B
8._____B
9._____A
10.____B
11.____C
12.____A
13.____A
14.____B
15.____A
16.____B
17.____C
18.____C
19.____C
20.____B
21.____B
22.____B
23.____C
24.____B
25.____B
26.____B
27.____A
28.____B
29.____A
30.____C
31.____B
32.____C
33.____C
34.____B
35.____C
36.____B
37.____C
38.____C
39.____B
40.____A
41.____D
42.____A
43.____B
44____B
45.____C

ANSWERS

5_Hazardous Materials Practice Test. Answers

1.____A
2.____B
3.____A
4.____A
5.____C
6.____A
7.____C
8.____B
9.____C
10.____A
11.____B
12.____C
13.____A
14.____B
15.____C
16.____A
17.____A
18.____A
19.____C
20.____C
21.____B
22.____A
23.____C
24.____A
25.____B
26.____B
27.____B
28.____C
29.____A
30.____A
31.____B
32.____B
33.____A
34.____A
35.____B
36.____B
37.____C
38.____B
39.____C
40.____B
41.____A
42.____A
43.____D
44.____C
45.____D
46.____B
47.____C
48.____D
49.____B
50.____C
51.____B

6_Double/Triple Trailers Practice Test. Answers

1.____A
2.____A
3.____B
4.____A
5.____C
6.____C
7.____A
8.____A
9.____B
10.____B
11.____A
12.____C
13.____A
14.____B
15.____D
16.____B
17.____C
18.____A
19.____B
20.____A
21.____A
22.____C
23.____C
24.____C
25.____B
26.____B
27.____C
28.____B
29.____A
30.____C
31.____C
32.____C
33.____B
34.____A
35.____B
36.____B
37.____D
38.____A
39.____D
40.____C
41.____D
42.____C
43.____B
44.____A
45.____B
46.____A

7_Combination Vehicles Practice Test. Answers

1._____A
2._____B
3._____A
4._____A
5._____B
6._____A
7._____B
8._____C
9._____B
10.____B
11.____A
12.____C
13.____B
14.____A
15.____B
16.____A
17.____A
18.____B
19.____A
20.____A
21.____B
22.____A
23.____A
24.____A
25.____B
26.____C
27.____B
28.____B
29.____C
30.____A
31.____A
32.____C
33.____C
34.____C
35.____C
36.____A
37.____C
38.____C
39.____A
40.____B
41.____A
42.____A
43.____C
44.____B
45.____C
46.____A
47.____B

8_School Bus Practice Test. Answers

1._____A
2._____B
3._____B
4._____B
5._____C
6._____C
7._____B
8._____B
9._____A
10.____D
11.____B
12.____B
13.____D
14.____A
15.____B
16.____B
17.____C
18.____B
19.____A
20.____C
21.____D
22.____B
23.____D
24.____A
25.____C
26.____C
27.____D
28.____D
29.____C
30.____A
31.____D
32.____A
33.____A
34.____C
35.____B
36.____C
37.____B
38.____D
39.____D
40.____D
41.____D
42.____A
43.____B
44.____A
45.____C
46.____B
47.____C
48.____B
49.____C
50.____B
51.____A
52.____B

CONCLUSION

Thank you so much for purchasing **"CDL Practice Test: Includes Full Length Exams for all Classes + Audiobook, Flashcards and Online Videos."**

There are so many resources on the market today for test prep and study materials, so we are proud that you have chosen us to help you through to the next part of your career phase.

This book was written with the utmost care for correct, complete, and factual information that can help you to pass your commercial driver's license exam the first time you take it.

Driving professionally is a wonderful career path that has the potential to take you all around the highways and routes of North America. Traveling on this nation's roads and seeing those areas of the country allows you to see the real beauty that this country has to offer!

It is our hope that the information in this book has enlightened you on the regulations, standards, and precedents in the automotive and transportation industries. These regulations were painstakingly created and put in place by those who did the research to find us the safest and most efficient means of regulating travel on this nation's highways.

For more information on these topics that is right from the source, please check the websites for the federal agencies listed in this book such as the TSA, the PDTI, the DOT, and more.

Thank you once again for reading and best of luck to you in your examinations and your new career as a commercial driver!

Thank you for reading This book!

If you enjoyed it, please visit the site where you purchased it and write a brief review.

Your feedback is important to me and will help other readers decide whether to read the book too.

LEARNIK
Press

Made in the USA
Las Vegas, NV
07 April 2025

20637909R00059